CHILDHOOD, FAMILY, ALCOHOL

T0270825

This lucid, engaging text provides a much-needed insight into drinking cultures and practices in families with younger children. Based upon an impressive programme of research, its key contribution is to demonstrate how alcohol is signified, articulated and felt, both within and beyond the confines of the family. The book should initiate a step-change in scholarship on childhood, families and alcohol, whilst providing clear recommendations for policy-makers working in this arena.

Peter Kraftl, University of Birmingham, UK

Jayne and Valentine's new book represents the cutting edge of research on alcohol studies. It is an excellent addition to a burgeoning area of research. The range and focus of the work is impressive and represents essential reading for anyone interested in understanding the complex worlds of children, families and alcohol.

Michael Leyshon, University of Exeter, UK

Childhood,Family, Alcohol

Cardiff University, UK

GILL VALENTINE
University of Sheffield, UK

LONDON AND NEW YORK

First published 2016 by Ashgate Publishing

2 Park Square, Milton Park, Abingdon, Oxfordshire OX14 4RN
52 Vanderbilt Avenue, New York, NY 10017

Routledge is an imprint of the Taylor & Francis Group, an informa business

First issued in paperback 2019

British Library Cataloguing in Publication Data
A catalogue record for this book is available from the British Library

The Library of Congress Cataloging-in-Publication Data has been applied for

ISBN 13: 978-1-138-41611-6 (hbk)
ISBN 13: 978-0-367-21905-5 (pbk)

Contents

List of Tables

Acknowledgements

We would like to thank the Joseph Rowntree Foundation for supporting the research presented in this book. In particular, we are very grateful to Charlie Lloyd and Claire Turner for their encouragement and advice throughout the project. We also want to acknowledge the Advisory Group for their valuable contributions to the development of the research: Betsy Thom (Middlesex University), Dianne Draper (Department for Health), Clem Henricson (Parenting and Family Institute), Anne Delargy (Alcohol Concern), David Foxcroft (Oxford Brookes University) and Jackie Marsh (University of Sheffield). Thanks also to Myles Gould for guiding the quantitative, and Julia Keenan for undertaking the qualitative elements of the study.

Finally, we would like to thank the families who participated in this study for sharing their experiences with us.

Mark would like to thank his colleagues at the University of Manchester for our decade together (2005–2015) as well as and also those 'down the road' at Manchester Metropolitan University, as well as Bethan Evans, David Bell, the Wollongong Boys, the Leyshon's, Phil Hubbard, Sarah L. Holloway and Dr and Mrs Potts. Special thanks to First Class Daisy.

The authors would like to thank Carolyn Court and Valerie Rose at Ashgate for their support and patience.

Title page photograph by Bethan Evans.

Some chapters in this book draw, in parts, on empirical material and arguments published elsewhere. The authors and publishers would like to thank copyright holders for permission to reproduce material as follows:

Valentine, G., Jayne, M. and Gould, M. (2014) 'The proximity effect: the role of the affective space of family life in shaping children's knowledge about alcohol and its social and health implications', *Childhood: a Journal of Global Child Research*, 21(1): 103–118. Reproduced by permission of

Sage Publications Ltd, London, Los Angeles, New Delhi, Singapore and Washington DC.
Jayne, M. and Valentine, G. (2015) "'It makes you go crazy": children's knowledge and experience of alcohol consumption', *Journal of Consumer Culture* (Online First). Reproduced by permission of Sage Publications Ltd, London, Los Angeles, New Delhi, Singapore and Washington DC.
Jayne, M., Valentine, G. and Gould, M. (2012) 'Family life and alcohol consumption: the transmission of "public" and "private" drinking cultures', *Drugs: Education, Prevention and Policy*, 19(3): 192–200. Reproduced by permission of Routledge, London and New York.
Valentine, G., Jayne, M. and Gould, M. (2011) 'Do as I say, not as I do: the affective space of family life and the transmission of drinking cultures', *Environment and Planning A*, 44(1): 776–792. Reproduced by permission of Pion Ltd., London and New York.

Chapter 1
Introduction

In contemporary societies in the global north there is growing concern about alcohol consumption amongst young people, even in countries such as France and Italy, which have previously been assumed to have 'sensible' drinking cultures (Järvinen and Room 2007; Velleman 2009). Alcohol research has generated a voluminous amount of writing focused on young people in terms of a diverse range of issues with a non-exhaustive list of studies including; everyday drinking practices in households and the role of family and peer influence (Komro et al. 2007; Lowe et al. 1993; Yu 2004; Conway et al. 2003; Shucksmith et al. 1997; Marquis 2004; Bogenschneider 2004; Plant and Miller 2007; Bergh et al. 2011); teenage 'risky' behaviour (Newburn and Shinner 2001); gendered geographies of young people's drinking (Forsyth and Bernard 2000; Hubbard 2005; Leyshon 2005, 2008) and the mis-use of alcohol by both young people and parents/carers (Leib et al. 2002; McKeganey et al. 2002; Ward and Snow 2010; Templeton et al. 2011). Despite this progress it is noticeable that there has been a relative lack of research that examines the transmission of drinking cultures within families across a broad diversity of social groups, including those who do 'not necessarily consider themselves as having an alcohol problem, or to be suffering the consequences of other people's problematic drinking' (Holloway et al. 208: 534). Moreover, research has also failed to consider in a sustained manner children who are younger than teenagers, or indeed to address in a convincing way how space and place are key constituents in parental and children's and young peoples' knowledges and experiences of alcohol, drinking and drunkenness.

In this book we unpack the complex ways in which geographical imaginations contribute to the transmission of drinking cultures within families (Jayne et al. 2008a; Valentine et al. 2010a). In the UK, official figures actually show a decrease in numbers of young people drinking above Government recommended levels of consumption, with politicians, policy makers and charities warning against complacency and raising concerns about young people underreporting their alcohol consumption and highlighting the role of alcohol in violence and disorder in public space (BBC 2015; and see Jayne and Valentine 2015; Jayne et al. 2015 for a critique). Recent policy attention has, for example, been focused on the potential role of parents/carers in preventing alcohol misuse by their offspring, and in supporting the introduction of alcohol to young people.

For example, in 2009, for the first time the UK Government published *Guidance on the Consumption of Alcohol by Children and Young People* in order to offer young people and their parents/carers advice on how to identify and prevent problem drinking (Donaldson 2009). In this report the then Chief Medical Officer, Sir Liam Donaldson, recommended that young people under the age of 15 should avoid alcohol completely, 15- to 17-year-olds should only consume alcohol with the guidance of a parent/carer, and certainly no more than once a week (Donaldson 2009). The former Chief Medical Officer's guidelines followed previous policy initiatives focused on young people such as the *Youth Alcohol Action Plan* (2008) which emerged from the UK government's *Alcohol Harm Reduction Strategy for England* (2004) and *Safe, Sensible and Social* (2007), which sought to address alcohol-related problems for individuals, families and communities. Indeed, the UK charity Alcohol Concern has also argued that the law concerning giving alcohol to children at home should be reviewed and called for a ban on the advertising of alcohol on television before the 9pm watershed and before non-18 films in cinemas, as well as for the National Curriculum to include increased alcohol education (Diment et al. 2007).

The government's published alcohol strategies have also been complemented by a national television advertising campaign entitled *Why Let Drink Decide?* where teenage actors describe their future 'risky and harmful' decision-making dominated by alcohol. As well as offering links to the advertisements, the campaign website provided advice about how to 'be in control', to 'think before you drink', to 'drink less', as well as guidance on 'looking after a drunken friend' and 'having fun without alcohol' (*Why Let Drink Decide?* 2010). These policy initiatives have helped ensure that the role of adults and families in transmitting values and practices relating to alcohol consumption to children has become a central feature of political and popular debate. Indeed, academics have long argued that 'the family is the primary context for the socialisation of drinking behaviour in young people' (Foxcroft and Lowe 1997: 227) and consequently a major influence on the development of the drinking careers of young people in relation to both drinking habits and attitudes to drinking (e.g. Raskin White et al. 1991).

Theorizing Childhood, Families, Alcohol

Despite there being a backdrop of academic debate, political and popular concern and policy focus the role of alcohol within the family has nonetheless been under theorised. For example, research has been dominated by quantitative analysis of the links between family structure and risk (e.g. socio-economic factors, parental alcohol problems etc.), with only a handful of studies addressing parents'/carers' communication and supervisory strategies (all with teenagers) (e.g. van der Vorst et al. 2005). As such, the overwhelming focus of research on family life can more accurately be understood as studies of social reproduction, comprising mainly research on child-care, and the organisation of domestic labour (e.g. Cox and Narula 2003; Dyck 1996; Holloway 1998, 1999). As such, they have been quite instrumental, focusing on parenting decisions rather than the banal everyday experiences of living in families (with children's experiences of family life located within distinct sub-fields, such as children's geographies, psychology, consumer culture and some 'alcohol studies' – which will be discussed throughout this section). There is therefore a lack of theoretical and empirical research focused on families across a broad diversity of social groups with children younger than teenagers, and therefore relatively little is known about whether parents/carers actively teach *pre-teen* children to drink within a family setting. As such, a recent review observed the need for more studies of parents'/carers' attitudes and practices in relation to children's alcohol consumption (Smith and Foxcroft 2009).

For example, it is through shared everyday practices and interactions that family relations and individuals' identities, attitudes and values are forged (Morgan 1996). As such, theorists and researchers need to pay more attention to what it means to live in families, how family connections are constituted and lived between people, to the unpredictable flow of daily events and inconsistencies of family behaviour, and to the relationships between the family in the present and the pasts/futures of its members. By foregrounding everyday life, and the negotiation of shared attitudes and practices in families (in relation to material objects, emotions, time etc.) rather than focusing on individuals within families or aggregate patterns in family behaviour, in this book we centralize the dynamic of family life with reference to social and cultural theory.

Gillis (1996) argues that everyone lives in two families – *the one they live by* (i.e. our idealized vision of family life which we aspire to) which serves as a moral anchor for the way we believe family life ought to be lived, and the *one they live with* (i.e. the families we share our everyday realities with) with all their contradictions, messiness and disorder. However, such understandings of 'family' tend to get lost in definitions used by policy-makers to describe households in which children live and Gillies (2009) argues that the State has been adopting an increasingly interventionist approach towards families as the mundane practices of everyday family life (e.g. eating) have been systematically linked with the health and well-being outcomes not only of 'the child' but also of society as a whole (obesity etc.). Whereas the nineteenth century was marked by the introduction of compulsory schooling in western societies to compensate for the perceived domestic deficiencies of 'working class' families, contemporary parents/carers are increasingly charged with responsibilities to maximize their child's start in life by providing the 'right' support at home as part of processes of individualization (Beck and Beck-Gernsheim 1995; Gagen 2000). Most notably, the contemporary family is perceived to be the site where our personhood is cultivated and the lens through which pasts, presents and futures are often interpreted, acted upon and imagined (Gilles 2009). Growing anxiety about children's well-being has therefore meant that many aspects of family life have become subject to government legislation, policies, and 'advice' (Plummer 2003).

It is within this broader context that a burgeoning body of literature by social and cultural theorists and more specifically writing focused on consumer culture has sought to highlight the complexities and contradictions of 'childhood' and 'family' life and to respond to the challenge that 'relatively little is known about how children engage in practices of consumption or the significance of this to their everyday lives and broader issues of social organisation' (Martens et al. 2004: 42). Recent writing for example has considered discursive constructions of children's consumption of nature/culture through children's oral, textural and visual representations of 'the city' and observation of children's urban spatial practices (Wells 2002). The role of homemade food in family identity has also been considered through practices that mark intergenerational care-giving, altruism and love as model 'characteristics' (Evans et al. 2011). Tyler

(2009) discusses sales-service environments and encounters between sales staff and children; and Peterson (2010) focuses on children's purchasing and playing of computer games and collecting of objects as an everyday social performance. This research identifies important signifiers of how parents, carers and other adults 'grapple with transmitting domestic cultural values, enacting roles and responsibilities of parental generations towards their other family members, and defending the domestic unit against the fissiparous pull of competing moral discourses' (Mosio et al. 2004: 379–38).

Underpinning this work is also a desire to uncover 'theoretical and empirical insights into the lived experience of young people as they mediate the shifting milieus of their social lives' (Martens et al. 2004: 42). Indeed, Ruckenstein (2010: 401) suggests that children, childhood and families are becoming increasingly important in rethinking and revising notions of consumer culture by adopting 'a child centred perspective on consumption [which] supports the study of temporal dimensions of consumption by emphasizing how consumer culture reproduces *and* transforms itself through the lifecycle and over generations'. This approach develops understanding of childhood and family as key to social formation of global consumers, and children as a major target market for global capital, acknowledging childhood 'enchantment' around consumption but avoiding developmentalist and sacralised theorization (Langer 2004). For example, Cook (2003: 150) argues that 'emergent from the interplay between adult and child is the iterative production of social space. [and that] Children, of course, do not passively accept impositions … [highlighting that] within childhood there exists a dynamic series of oppositional stances defining children of different age-genders in relation to each other'. Valentine's (1999: 21) contention nonetheless that it is adult power which defines 'what is right and wrong (safe and unsafe)' has led to theoretical and empirical reflection on the ways in which while 'the individual might be the bearer of anxiety … it must ultimately be the collective social valuing and discursive formation of 'competent parenting; that underlie any anxieties about parenting' (Martens et al. 2004: 168).

In reviewing these debates, Cook (2008: 233) argues that critique from studies of children's consumption highlights that (and we would suggest the majority of researchers focused on childhood, family and alcohol replicate is) a 'socialization perspective, is at base, teleological in its epistemology in the sense that it posits children as incomplete, less-than-knowledgeable persons whose movement is towards an assumed or desired state of being and knowing'.

Moreover, Cook (2008: 222) goes on to suggest it is important that children are considered as important actors in order 'to disrupt individualistic assumption about economic action by bringing women, mothers and caregivers into the picture … [so that the] relational and co-productive nature of acquiring, having and displaying things becomes necessary, evident and unavoidable'. This highlights the need to understand unequal power relations between different social groups, as well as interrogating processes and practices of learning and associated lifestyle and identity formation.

More specifically, Martens et al. (2004) point to the significant mileage in understanding children's engagement with material culture *and* parent-child relationship via networks of adults who play a role in children's lives. Such comments notwithstanding, Martens et al. (2004: 175) suggest it is important that research agendas acknowledge children's consumption is intimately connected to parents'/carers' consumption (particularly as parents/carers often consume on behalf of their children) and as such there is a need to understand 'how children 'learn' to consume, the lifestyles of their parents/ carers, the ways that their parent/carer reflexively engage with memories of their own childhood (or biography) and parental readings of material culture all lie at the heart of what can be understood as children's consumption'. Writing on consumer culture, childhood, and families has thus begun to productively engage with theories 'about the individuality of desire, identity and lifestyle … [as being discursively and differentially constructed through] relationships, obligations and reciprocity' (Cook 2004: 237).

Social reproduction, adult-children interaction, materialities and intergenerational transmission of consumption cultures are thus central to theoretical understanding of childhood, families and alcohol, drinking and drunkenness. Indeed, while being a relatively small body of writing within the voluminous 'alcohol studies' literature, for almost 40 years a handful of progressive theorists across the social sciences have considered alcohol in the lives of children and families in different countries around the world in a similar way. For example, Ritson (1975) identified how children distinguish between alcoholic and other drinks and develop the view that drinking is a mysterious adult activity that is taboo for children. Ritson concluded that the significance that children attach to alcohol and their own intended future drinking relates to the drinking habits of their family. Ritson utilized methods such as smell recognition, which highlighted indifference of children aged 6–10 to trying alcohol for themselves, and the ways in which interplay of parent's/carer's attitudes, peers, advertising and social factors combine to influence children's understanding of adult drinking.

Ritson pointed to the ways in which alcohol becomes a powerful symbol of adulthood and adult sociability for children aged 10 years old. However, nuancing this line of argument, Webb et al. (1996) suggests that peer, parental and children's own attitudes towards alcohol are linked to intentions to drink in fifth grade, whereas two years later it is peer influence that increasingly is an important factor. Also, following Ritson's (1975) study into materialities of children's understanding of alcohol, drinking and drunkenness, Noll et al. (1999) investigated whether children aged 30–72 months can identify alcoholic beverage by smell and identified a positive/negative link of specific drinks to parental alcohol consumption and behaviour. In a similar vein, other research highlights how children consider particular drinks as having positive or negative effects on adult behaviour, for example, champagne and wine, as opposed to beer (Gaines et al. 1988; Casswell et al. 1985).

More recently, Ward et al.'s (2010) research in Australia and New Zealand focused on parental concern to ensure 'supervised introduction' of alcohol to their children. On a similar tack, Rolando et al. (2012) highlights how children get acquainted with alcohol in socially, culturally and geographically diverse ways. Research in Italy and Finland, by Rolando et al. (2012) suggest that in the former, positive values of family life are typical of first experiences of under-age drinking whereas in Finland first experiences tend to relate to intoxication with peers. Such research points to social reproduction of 'national drinking cultures' (see Jayne et al. 2008 for a critique). In contrast, as highlighted at the beginning of this chapter Jarvinen and Room (2007) point to changing patterns of 'national drinking cultures' showing that in France and Italy, which have previously been assumed to have 'sensible family drinking cultures', there is growing concern over young people 'binge drinking'.

Further work unpacking alcohol socialization (monitoring of parental alcohol use by children) in the UK, suggests that attitudes of children aged 5–10 years old towards alcohol consumption had changed little over the previous 20 years, although arguing against past research to suggest that children have an overwhelmingly negative attitude to adult drinking (Jackson et al. 1997; Fossey 1993). In contrast Eadie et al. (2010) suggest that children aged 7–12 have sophisticated knowledge's and understanding of alcohol and are able to identify differing levels of adult's intoxication and differences between occasional and habitual drunkenness. Eadie et al. (2010) also argue that 'the home' is an important site of everyday alcohol use, and how special occasions are explicitly marked by alcohol consumption, which engenders positive family experiences for children. However, Eadie et al. (2010) highlight that rather than teaching their children about alcohol, parents/carers did offer advice and guidance towards smoking and drug taking. In reflecting on notions of social reproduction and transmission of drinking cultures, Ward et al. (2010) and Ward and Snow (2008) nonetheless critique research which suggests a causal link between alcohol related problems in later life extrapolated from experiences in childhood.

In reviewing the strengths and weaknesses of alcohol research related to children, young people and families, Velleman (2009) describes a genealogy organised around substantive empirical topics including: family processes and structures (parenting styles, family cohesion, sibling behaviour); peer selection (although little consensus); direct (advertising) and depictions (media representations, product placement etc.) related to marketing and cultural representations of alcohol; and country, ethnicity, religion, socio-economic status and other cultural factors. It is disappointing however that within this large body of writing that studies engaging with the complexity and 'messiness' of everyday life through theoretically grounded empirical research has not had a significant impact on debate and research agendas. Moreover, as Valentine et al. (2010), Ward et al. (2010) and Jackson et al. (1997) suggest there has also been an absence of longitudinal research which would enable more sophisticated understanding of the role of alcohol, drinking and drunkenness across the life-course. As such, while 'alcohol studies' have addressed similar theoretical and empirical terrain to theorists interested in social and cultural geography, psychology, consumer culture and so on, there has been little useful dialogue between these bodies of literature in order to respond to Cody's (2012: 41) challenge to address the 'dearth of both empirical and theoretical accounts of young adolescents' specific consumption practices as they attempt to mediate the intricacies of their lived experiences and social contexts'.

Data and Methods

To that end, the research presented in this book was undertaken in the UK and included a national survey of 2089 families with at least one child aged 5 to 12, and in-depth multi-stage qualitative research with 10 case study families (see appendices). The research methodology was designed in order to examine both *'extra-familial* 'norms' about parenting and drinking cultures which are generated in wider society (through: law/regulation; media/advertising, social networks etc.); and *intra-familial* 'norms' in terms of the attitudes to alcohol/personal drinking habits which parents/carers model through their own behaviour or the rules/advice they establish for their children at home (see Valentine et al. 2010a for more detail).

The first stage of the research involved a survey to collect data on parents'/carers' perceived (above/below/at recommended levels) and actual alcohol consumption practices; their perceptions of national/local norms in relation to attitudes towards the role of alcohol in the family; and their awareness of the law and perceptions of national/local 'norms' in relation to children and alcohol. The sampling strategy allowed us to establish national patterns in relation to parents'/carers' attitudes and practices towards the role of alcohol within the family, and to benchmark the qualitative case-study research within this national context. It was also used to recruit family participants for the qualitative research phases.

Ten families, with at least one child aged 5–12 were recruited via the survey as case studies for the multi-stage qualitative research (see Table 1.1). The case studies were purposively sampled on the basis of the survey results to include families with diverse structures, socio-economic profiles and a range of attitudes and practices to drinking. Interviews with parent(s)/carer(s) collected data about their attitudes/practices towards parenting, specifically in relation to alcohol in the family. Where families were constituted by two parents/carers they were interviewed together where possible. The children's experiences of the above issues were explored through a child-centred interview process that as well as 'conventional' interview-style questions included: exploring children's understanding of alcohol by asking them to identify samples of drinks (alcoholic and non-alcoholic) by smell; and asking children which drinks they could identify from a series of advertisements for common products/brands. The role of alcohol in the family was then explored with the youngest children by using puppets or a dolls' house to play-act a family party. Older children were shown clips from episodes of the cartoon series The Simpsons which represent both adults and children as drinking/drunk. These were used as a basis for a wider discussion about their attitudes to alcohol and family practices. In addition, the case study families were asked to invite a member of the research team to a family event where alcohol was consumed (e.g. birthday/anniversary/fireworks party, wedding etc.). A researcher also accompanied the families on a 'normal' treat that involved alcohol (e.g. meal out; a sporting/leisure/entertainment event; shopping etc.). This participant observation allowed the research team to build both descriptive observations and narrative accounts about children's interaction with adults in relation to alcohol in different family time-spaces.

Fieldwork diaries recorded after these activities allowed the researchers to build descriptive observations and narrative accounts about the children's interaction with adults in relation to alcohol in different family time-spaces. Such approaches are not replicable or generalisable given the presence of a researcher. While reflexive about the dynamics of these encounters we follow Rose (1997) in acknowledging that the complex nature of the multiple positionings and (dis)identifications produced in research encounters mean the influence of the researcher is always unknowable (see also Valentine 2002). Rather, as Gabb (2010) argues, what is of importance is the way the participants' represent themselves regardless of whether these performances are staged for the researcher or everyday acts. The interviews were recorded and transcribed, and along with the participant observation material were analysed using conventional qualitative techniques. The quotations included throughout the remaining chapters are verbatim. Three ellipsis dots indicate minor edits have been made to clarify their readability. The phrase [edit] is used to signify a section of text has been removed. All the names used are pseudonyms.

Table 1.1   Characteristics of the families participating in the case study element of the research (anonymised)

Family pseudonym Family type location religion	Family members marital status age/gender/ ethnicity	Highest earner's occupation, highest qualification	Children's drinking as reported by parents	Parental drinking level, frequency and amount
A: 'Atkinson' Nuclear North West England Christian	Father, aged 40s (Q respondent) Mother, aged 40s Married, living with partner Two daughters, aged 9 and 6 Father: White English Rest of family: White British	Father: Higher professional, degree Mother: Works part time, degree	Both daughters tasted and had a sip of wine at family celebration or home	Father: At/below recommended levels, never/ little Mother: Drinks less, less frequently
B: 'Bilton' Nuclear North West England No religion	Mother, aged 30s (Q respondent) Father, aged 40s Married, living with partner Son, aged 11 Daughter, aged 7 Mother: White English Rest of family: White British	Father: Intermediate, other qualification Mother: Intermediate, other qualification	Both children have tasted and sipped beer/lager/cider and wine at family celebration or home Eldest told off for drinking alcohol Children have seen parent(s) hungover	Mother: At/below recommended levels, two to three times a week Father: Drinks less than mother, less frequently Neither parent drinks in the working week, but 'binge' at weekends
C: 'Clough' Nuclear Yorkshire and Humber Christian	Mother, aged 40s (Q respondent) Father, aged 50s Married, living with partner Two daughters, aged 17 and 10 Son, aged 10 (twin of daughter above) Mother: White English Rest of family: White English	Mother: Intermediate, degree Father: Made redundant		Mother: At/below recommended levels, four or more times a week Father: Drinks at a similar frequency, although drinks less

Table 1.1   *Continued*

D: 'Durham' Reconstituted South East England Christian	Mother, aged 30s (Q respondent) Father, aged 30s Married, living with partner Son, aged 9 (different biological father, no contact relationship with child) Two daughters, aged 5 and 1 Mother: White English Rest of family: White English	Father: Intermediate, O-Levels Mother: Took redundancy, previously worked in sales	All children have tasted alcohol, elder two have had sips of beer/cider/lager and/or wine at home and on holiday Children have seen parent(s) with hangovers	Mother: Above recommended levels, four or more times a week Father: Also drinks above recommended levels
E: 'Ellis' North West England No religion	Mother, aged 40s (Q respondent) Father, aged 40s Married, living with partner Son, aged 9 Mother: White English Rest of family: White English	Father: Intermediate, O-Levels Mother: Full-time parent, registered disabled	Child hasn't tasted alcohol, as "is too young"	Mother: Above recommended levels, four or more times a week Father: Drinks less alcohol and less frequently
F: 'Fisher' Reconstituted South East England Atheist	Mother, aged 30s (Q respondent) Father, aged 30s Living with partner/cohabiting Two sons, aged 16 and 14 (from father's previous relationship) Two daughters, aged 6 (from mother's previous marriage) and aged 4 months (biological daughter) Mother: White English Rest of family: White English	Mother: Lower managerial/ professional (part-time), higher degree Father: Self-employed businessman, A-levels	Daughter aged 6 has had a taste and sip of wine (and fortified wine with mixer) beer/cider/lager, champagne/cava	Mother: At/below recommended levels, monthly or less Father: Drinks more, more frequently, but doesn't get drunk
G: 'Green' Nuclear North East England	Mother, aged 50s (Q respondent) Father, aged 50s Two sons, aged 14 and 10 (twin, see below) Two daughters, aged 10 (twin of son) and 8 Mother: White Irish (self-assigned) Father: White British	Mother: Intermediate Father: Intermediate	Eldest son told off for drinking alcohol	Mother: Recommended levels, two to three times a week

Table 1.1   *Concluded*

H: 'Harris- Bailey' Nuclear Yorkshire and Humber No religion Mother, aged 30s (Q respondent) Father, aged 40s Living with partner/cohabiting Three daughters, aged 7, 20 months and 4 months Mother: White English Rest of family: White English	Mother: Lower managerial and professional, degree Father: Lower managerial and professional (short-term contract)	Eldest daughter has had a taste and sip of alcoholic drinks, and a watered down drink: wine, beer/lager/cider and champagne/cava at home, family event, parents' friend's home in a pub and at a public festival Eldest daughter has possibly seen mother with a hangover	Mother: At/below recommended levels, four or more times a week Father: Similar to mother
I: Irwin Single-parent Yorkshire and Humber Christian (Catholic) Mother, aged 40s Single Participant's mother, aged 70s, also resides with family on a semi-permanent basis Two daughters with different fathers, aged 7 (sees father every Saturday) and 4 (sees father regularly on informally)	Lower managerial/ professional, higher degree	Eldest daughter has had a taste and a sip of beer/cider/lager – given to her by her biological father at the pub	At/below recommended levels, monthly or less
J: Jones Reconstituted, elder children elsewhere Yorkshire and Humber No religion Mother, aged 40s (Q respondent) Father, aged 40s Living with partner/cohabiting Daughter, aged 7 Two elder children from previous relationship (over 18) now living independently elsewhere	Mother: Lower managerial/ professional, A-Levels	Daughter has not had alcohol to mother's knowledge – is not allowed it and child dislikes the idea of it	At/below recommended levels

Source: Family Life and Alcohol Consumption questionnaire survey (2010a)

Reading Childhood, Families, Alcohol

In this book we outline the following contributions to the advancement of understanding of childhood, families and alcohol. Firstly, we highlight the importance of undertaking a cross-generational perspective by exploring pre-teen children's understandings of alcohol as well as that of parents/carers and other adults. This is because adults and children may experience familial socialisation practices around alcohol differently. Too often adults' views about what is in the best interests of children are read through the lens of age-appropriate behaviours which are predicated on deterministic theories of child development in which pre-teens are presumed to be too immature to express opinions, rather than on children's own experiences of their life-worlds. Drawing on theoretical understandings from social studies of childhood, we recognise that children are agents in their own worlds and active choice-makers in terms of consumption, and lifestyle/health behaviours (e.g. Alderson 1993). Indeed, it is through the negotiation of shared practices that individual identities (of parents/carers and children), and family relations are forged (Morgan 1996). Consequently, to understand the place of alcohol in children's lives we need to pay attention to how families are lived between people, to daily events and the inconsistencies of family behaviour (Valentine et al. 2012). As such, we focus on young children's own knowledge about alcohol, and its role within the context of their family lives.

Secondly, our approach is concerned with the significance of the spaces of everyday family life: understanding childhood in context (Valentine and Hughes 2012). Traditional research on child development has commonly looked at the child in isolation in which behaviour is assumed to reside in the subject. Rather, our work is broadly informed by insights from Lev Vygotsky (1978), a Russian psychologist, who explored the role of culture and interpersonal interaction in child development. He theorised that children copy significant others (particularly parents/carers but also other adults/peers) and in doing so advance their learning beyond what they would achieve independently. His 'zone of proximal development' has been investigated empirically by comparing individual with assisted problem solving tasks. More broadly, his educational development theory demonstrates the socially embedded nature of learning. It is through interactions in everyday spaces, like the home, that children derive meaning and gain shared cultural knowledge including the taken for granted social rules of behaviour that often pass unnoticed in everyday life. While not strictly taking a Vygotskyian perspective, we focus on the proximity effect – the way that nearness in space and time can bring with it a sense of identification – to highlight the significance of the experiential (e.g. habitual routines and shared practices) in terms of the way that families are created and lived together in shaping children's knowledge about alcohol. In doing so, we contribute towards addressing Daly's (2003) concern that what it means to live in families remains an elusive challenge for social scientists.

Thirdly, we argue the importance of considering adult alcohol consumption, drinking practices and drunken performativities in relation to children, childhood and family with reference to social/individualized consumption, social reproduction, adult-children interaction, materialities, and intergenerational 'transmission' of drinking cultures. Such an approach is important contribution to debate across the social sciences and theoretical and empirical moves away from developmentalist notions of children as 'yet-to-be adults', towards consideration of lived experiences and social and spatial contexts of how drinking cultures are reproduced through the life-course and across generations (Evans 2008). As Cook (2003, 2008) argues, there is a need to disrupt teleological depictions of childhood/adulthood and notions that children passively accept impositions, so that dominant notions of socialization as individualistic can be challenged. This perspective foregrounds relationships, obligations and reciprocity bound up with consumption practices and identity formation in order that everyday lives of children and adults can be better understood.

Finally, we advance work that has focused on bodies and embodiment and non- (and more-than) representational understanding of complexities of childhood and family life (Horton and Kraftl 2006; Evans 2010; Kraftl 2013). Key arguments suggest that emotions reside in bodies and places and exist as relational flows, fluxes and currents, in-between people and places. Theorists highlight bodily boundaries are frequently transgressed in emotionally powerful, disruptive and conflictual ways (Longhurst 2001). Research into joy, sadness, fear, love, hate etc. has highlighted how emotions matter. As Davidson et al. (2005: 10) suggest, our lives can be bright, dull or darkened by our emotional outlook and unpacking emotions (experientially and conceptually) must be understood with reference to 'socio-spatial mediation and articulation rather than as entirely interiorised subjective mental states'. Such writing advances understanding of fluidity of boundaries between emotions, embodiment and affect. Venn (2010: 132) highlights 'mind-body-world'; where human existence is characterized by instantaneous correlation of 'information': facts, signals, rumours, news, mixed in with moods and emotional energies, enables agents to participate in an activity in which they behave both as an individual and as an element of a collectivity'. Throughout this book we highlight how emotions, embodiment and affect add value to theoretical and empirical agendas regarding childhood, familes and alcohol, drinking and drunkenness (Jayne et al. 2010; Valentine et al. 2013).

We apply these four theoretical bodies of work to arguments and empirical evidence presented throughout the remaining chapters. Chapter 2 focuses on the transmission of attitudes and practices relating to alcohol in everyday family life. In doing so, we focus on parents'/carers' ideals of how family life and alcohol consumption should be – in Gillis' terms *'the families we live by'* and also explore the often taken for granted realities of *'the families we live with'* in order to understand the complexities of how parents/carers model different attitudes towards alcohol. Chapter 3 considers the 'rules and guidance' bound up with transmission of drinking cultures within families. We highlight the differential and discursive construction of the home as a space where parents/carers are happy to introduce children to alcohol in a 'safe' environment in opposition to public spaces which they consider to be locations where alcohol consumption is associated with violence and disorder. Chapter 4 focuses on children's knowledge and understanding about alcohol, addressing three questions: what do pre-teen children know about alcohol and how do they learn about alcohol and its associated potential harms? Chapter 5 highlights how consideration of intergenerational transmission of practices relating to alcohol, drinking and drunkenness can contribute to theoretical and empirical debates relating to social reproduction, adult-children interaction, materialities and intergenerational transmission of drinking cultures that are bound up with social rather than individualized notions of consumption. By adding non-representational theories relating to emotions, embodiment and affect to such an agenda we also point to new fruitful avenues for research on childhood, families and alcohol. In the concluding chapter we consider the implications of our research findings for policy and practice and review the impact of our work beyond the academy.

Chapter 2
Attitudes and Practices

In this chapter we focus on the transmission of attitudes and practices relating to alcohol in everyday family life. In doing so, we focus on parents'/carers' ideals of how family life and alcohol consumption should be – in Gillis' terms *'the families we live by'* and also explore the often taken for granted realities of *'the families we live with'* – in order to understand the complexities and contradictions of the ways in which parents, carers and adults model attitudes and practices about alcohol to children.

The Families We Live By: Parents'/Carers' Perceptions of How Alcohol Ought to be Introduced to Children

Our research suggests that the family is a significant influence on the timing of young people's first consumption of alcohol. The current legal age at which children in the UK can consume alcohol in the home is five years old, yet the parents/carers who responded to the survey claimed that children should be in their mid-teens before it is acceptable for them to experiment with alcohol at home, and an adult before they should drink unsupervised in public space (see Table 2.1). These parental attitudes correspond quite closely with the advice from the former Chief Medical Officer, that children under the age of 15 should avoid alcohol completely, and that 15–17 year olds should only consume alcohol with the guidance of a parent. Similar research in New Zealand has produced an even more conservative response from parents/carers, with the majority of respondents to a survey identifying 18 as the average age at which young people should start drinking (Kypri et al. 2007). There were some minor differences in the attitudes of the UK parents/carers we surveyed in terms of social class, levels of family support and whether parents/carers drink above weekly recommended levels of units (see Appendix 3). Specifically, the average age which parents/carers thought it acceptable for their children to have a watered down alcoholic drink within a family meal and to fetch, pour or serve alcoholic drinks for others at home, for parents/carers who never drink alcohol is 16. In contrast, for those who drink at/below recommended levels it is 14, and for those who drink above recommended levels the figure was much lower at 12.

Table 2.1   Average age at which respondents think it is acceptable for children to engage in alcohol related activities

Acceptable for children to ...	Average age
Be given a taste of an alcoholic drink	13
Fetch, pour or serve alcoholic drinks for others at home	14
Have a watered down alcoholic drink with a family meal	14
Get away with having a sneaky sip of an alcoholic drink at a family event	15
Have an alcoholic drink at family events	16
Have an alcoholic drink with just their friends, in the family home under supervision	16
Drink unsupervised with their friends on Friday/Saturday nights	18

Source: Family Life and Alcohol Consumption questionnaire survey (2010a)

The parents/carers who participated in the qualitative research nonetheless had very diverse introductions to alcohol in their families that ranged from positive role models through to negative experiences of parental alcoholism. The interviewees who described their parents/carers as positive role models recalled being introduced to alcohol as a normal, acceptable, taken for granted part of family life. The memories capture the sociality and pleasures of drinking with family meals and special occasions. Being allowed to participate in this drinking culture – either by mimicking adult practices with a non-alcoholic drink or by actually being allowed alcohol – was remembered by interviewees as an important ritual in marking their gradual transition from childhood to adulthood:

My first drink I can remember vividly ... we were allowed, me and ... my older sister, we were allowed one Babycham on Christmas Day. And I think my parents started this when I was about 13 and that was like the highlight of Christmas Day forever, my one Babycham (laughs). [edit] It is such a memory, it was just like the best thing ever! [edit] So that was my … not my first drink ever; I think I'd had sips of drinks that my parents had given me but that was my first ever, my glass, my drink.

(Mother, Family E)

Yet, not all of the interviewees had such positive introductions to alcohol. Some had unpleasant childhood experiences because there was a heavy drinker/alcoholic in the family, or they remembered occasions spoiled because of alcohol. These negative memories made them determined to adopt a very different approach to both alcohol and parenting with their own children:

I remember the Christmas we found out that Father Christmas wasn't real … I was about 8 or 9 and I remember waking … I heard a crash on Christmas Eve and me and my brother woke up, went onto the landing and my mother had gone up into the attic to get the Christmas presents out and because she was drunk, she'd misplaced her foot, fell out of the attic and she was absolutely … she was so drunk, she really was. And that's how we found out Father Christmas didn't exist, by my drunken mother [Edit] She wasn't a doting parent. Whereas now, I hug ours all the time … and I them I love them and stuff like that [edit]. The main difference is between my childhood and … my kids are … are much more confident with the knowledge that they are loved and they're looked after … I have routines for them … There were no routines for us … never knowing were they going to be in? … Whereas for my two, it's just completely different. I don't hit them … [edit] I'll sit them down and I'll ask them what's happened and I'll explain to them when something's right or wrong. And I think that's a big thing, explaining things and making sure that they know that they're loved.

(Mother, Family B)

caring for small children … I can't imagine it with a hangover because when you're tired anyway, you end up shouting at them. I can see why my Dad with a hangover was a horrendous parent because it just doesn't go, you know. If you're tired anyway and then you've got four children [edit] … with needs you know, 'I want this', 'I want that' you know, how are you going to manage that with a hangover? … And it's either your kids are going to suffer or you're all going to suffer.

(Mother, Family I)

Other interviewees also reflected critically on the way they were parented for reasons that were not related to alcohol. In particular, these respondents felt that their own parents/carers had been over-protective/controlling or that their childhood family relationships had not been open and honest because their own parents/carers had been distant and authoritarian. Consequently, as children these respondents had not felt able to confide in their parents/carers about personal issues such as alcohol, drugs and sex, or that their parents/carers had kept secrets from them. Like the quote above describing childhoods as 'blighted' by heavy drinking by family members, these interviewees despite having more positive experiences of family life, also wanted to avoid reproducing some perceived failing of their own parents/carers in their relationships with their own children. They too were seeking to develop close and loving relationships with their children by establishing practices of openness within their families that included directness about alcohol.

In particular, most parents/carers expressed concern that being too strict with their children might prove counter-productive. Often drawing on memories of their own childhood experiences as a resource, these parents/carers argued that issues which are made taboo automatically become more attractive to young people because the process of making the transition from childhood to adulthood necessarily involves testing parental boundaries, and engaging in risky behaviour is one way of demonstrating independence (c.f. Bodgenschneider et al. 1998). Their perceptions are supported by US research that suggests that 'controlling' approaches to alcohol and drugs by parents/carers are less effective than strategies involving open negotiation (Highet 2005). For example, a US study of 537 teenagers and parents/carers interviewed by phone twice over a yearly interval found that the young people responded to parental rules and admonishments about alcohol and tobacco by increasing their consumption in acts of rebelliousness (Ennett et al. 2001).

For most of the parents/carers who participated in this study the emphasis is on communication instead of establishing rules to shape their children's behaviour (this will be returned to in Chapter 3). Implicit in their accounts is a belief that children do not just accept the values or attitudes of older generations but rather have to experience things for themselves – and therefore necessarily to make their own mistakes – in order to have a sense of 'being themselves'. The parents/carers also recognised that as their children become teenagers and then adults they will have less and less influence over them. Nonetheless, by developing a close and open relationship with their children these parents/carers believe that they can equip them with the skills to make the 'right' individual choices about issues such as alcohol, while also creating an emotional bond with them which will enable their offspring to draw on their support if, and when, they have a bad experience. In this sense, they are adopting a neoliberal style of parenting, emphasising individualism (e.g. Gullestad and Segalen 1997):

> Mother: … if she says to me 'Oh can I have a sip, can I have a little taste?' … If I said to her 'No, you can't have that', course … she would be in the fridge … she would have done it by now if she was denied being able to have some. You know, it's kind of like well why are you having it? And it's like if I said 'Oh well it's really bad and you can't have that'
> Father: So by making it ordinary you completely withdraw any of its attractions to them as children for using it to kick against you …
>
> (Parents, Family H)

Here, in developing connections between their own childhoods and their parenting style, interviewees tended to reflect in most detail on the trajectories of their own drinking careers in order to picture their own children's lives. In doing so, parents/carers expressed the hope that their children would follow their own patterns of relatively positive experiences of youthful excess followed by settling down into what they perceive as more sensible consumption habits.

Reflecting on the attitudes about alcohol that they are attempting to convey to their own children, all of the case study parents/carers stressed the importance of moderation. All the interviewees enjoy the positive aspects of drinking themselves (as a source of relaxation, sociality and so on) to varying degrees and wanted their children to grow up to enjoy these pleasures too. At the same time, the parents/carers recognised the potential risks of drinking to excess – even if they occasionally did so themselves – and wanted to ensure that their children are aware of these dangers. Indeed, one family whose daughter is currently anti-alcohol were concerned that as a result when she becomes a teenager her views may swing to the other extreme and she may be vulnerable to drinking to excess. As such, they too shared the views of other parents/carers that children should be taught to appreciate alcohol so that they will in turn learn to consume it in moderation as adults:

> Father: I mean personally I'd hope that she'd have just a very sensible attitude towards alcohol. [edit] … Well you know, that it's not good in itself or bad in itself but that you know, it's to be kind of drunk in moderation I suppose. There are dangers in terms of excessive drinking and you know, that's kind of all there to be seen around. But it's not that she shouldn't touch alcohol you know, from the other side … I suppose a pragmatic outlook is what I'd hope that she'll develop.
> Mother: She can enjoy an occasional drink …
> Father: Yes, that's right … In terms of with Lucy I think in that sense, I think that her present attitude [she is anti-alcohol] is a bit over the top and in that sense, it wouldn't be right if she still had that attitude in five years' time really because you know, there's probably more kind of danger that she'd go the other way completely. I think it's good for children to have an educated view of drugs generally you know, and an awareness of what the dangers are, as well as you know, possible benefits.
>
> (Parents, Family J)

Most of the interviewees argued that the family was the main site where children should learn about alcohol rather than school. While some interviewees noted the potential role of education in reinforcing messages about responsible drinking and meeting the needs of those who do not get support at home, most parents/carers felt that their children needed tailored advice and guidance which they were best placed to offer because of their intimate knowledge of their children, their friends and the spaces that they inhabited and consequently the situations they were likely to encounter. In this sense, the interviewees continually re-inscribed an individualised model of parenting.

In this section, we have reflected on parents'/carers' idealized notions of how children ought to be introduced to alcohol. Here, most of our respondents drew on stories about their own childhood family lives as a resource (to be repeated or reacted against) to explain what they believe about how children make the transition to adulthood, how they perceive as parents/carers they should create their own family relationships, and how they want their own children to embody their attitudes/practices towards alcohol in their own future lives. Past family lives become anchor points that help to shape present and future families conceptually (i.e. in terms of the idealized family relationships individuals seek to create) even when these experiences have been negative or ambiguous. In this sense, it is difficult to think about the families we live by without implicitly acknowledging their connections with instantiations of families we have lived with – family is an iterative process. However, while the findings from the quantitative element of this research suggest that the dominant parental attitude (i.e. their perception of what ought to be the *extra-familial* social norm) is that children in general should not be introduced to alcohol at home until their mid-teens and ought *not* to be allowed to drink in public space until they reach adulthood. In the following section we consider what parents/carers actually do in practice (i.e. their *intra-familial* norms) within the families they live with.

Families We Live With: The Unintentional Transmission of Attitudes and Practices Through 'Modelling' in Everyday Life

Research suggests that having alcohol available at home is related to higher drinking levels amongst teenagers (van Zundert et al. 2006). Yet, despite this alcohol is a taken for granted or unremarkable feature of everyday household life (Holloway et al. 2008). Of the respondents to our survey: 54.8 per cent always/usually had alcohol stored in the home, 28.3 per cent openly displayed their alcohol in wine racks/ on sideboard/kitchen units, 33.6 per cent keep it in the fridge, 37.6 per cent in cupboards/cabinets, and 24.5 per cent in cellars/garage/under the stairs. Only 13.0 per cent of the respondents stated that they kept alcohol in locked cupboards and 16.6 per cent that it was stored on high shelves/cupboards out of reach of their children.

The majority of the parents/carers surveyed agreed (strongly or slightly) that it is acceptable for parents/carers to drink in moderation in front of their children (86.1 per cent), and that it is acceptable for parents/carers to have a drink with a meal in front of their children (86.0 per cent). For many of the interviewees drinking at home is an important form of relaxation, particularly because as parents/carers the responsibilities of child-care limit (or at least complicate) their opportunities to unwind by pursuing leisure activities in public space. Moreover, the days of the week that male respondents to the survey described drinking most are weekends (Saturday – 43.4 per cent;

Fridays – 24.6 per cent and Sunday – 16.6 per cent) when they could escape the pressures of work, yet this is also family time. Most parents/carers therefore appear to be modelling alcohol consumption to their children as a pleasurable leisure activity or treat which is a counter to the stresses of everyday work life. In some cases the interviewees also imagined enrolling their children into this family relaxation ritual:

> Mother: You don't have to go out, you don't have to bloody get in the car, you don't have to arrange a babysitter and although there are plenty more pleasures to have in life, they take a lot more effort than unscrewing a bottle of bloody wine or opening a bottle of beer to wind down, you know. Equally, Nigel will go to the cinema, he's got an unlimited card and I go swimming … But it's [alcohol] an instant wind down pleasure and it's there isn't it; I think that's a lot of the appeal is that it's … you can have pleasure out of something that doesn't take any time to do. Whereas I enjoy going swimming … but you've got to get in the car and drive there.
>
> (Mother, Family H)

> Mother: [referring to alcohol] A treat, yes. Yeah, it's a family thing, it's a family treat, a meal, a celebration, whatever … when I was allowed my first, I must have been about 13 … I can certainly picture him [son] in the right environment having a beer with us or certainly by the time he's 16.
>
> (Mother, Family E)

In this sense, although emotions such as pleasure and relaxation are embodied in individual family members they can also be transmitted through such modelling practices to others and experienced in terms of a collective family atmosphere. Just under half of the parents/carers (47.5 per cent) surveyed said that over the last 12 months their children had seen them offer a drink to make visitors feel welcome in the home. For most of the case study respondents' alcohol played an integral part in family get-togethers: a practice of hospitality of which the children were well aware. Most of the children described in their interviews how friends and relatives visiting their homes bring alcohol with them, and how their parents/carers take alcohol as a present when they are going to others' homes:

Well when we go on holidays to my Granny's house, my Mum and my granny have some wine. And sometimes in the holidays when Granny comes here, they have some wine as well. My Mum would usually bring wine to Granny's, as a present.

(Aileen, aged 8, Family G)

Yet, while parents/carers were relaxed about modelling attitudes and practices about their own home drinking they expressed concerns about the practices to which their children might be exposed in other people's homes. The emergence of non-traditional family forms has led some pessimistic commentators to suggest that the breakdown of the conventional nuclear family will lead to a disintegration of moral frameworks. However, the evidence of this study is that this is not so. Family F is a reconstituted family with both parents/carers having children from previous relationships, as well as a daughter of their own. The father's two eldest children live for two weeks in Family F and then two weeks at his ex-partner's (their biological mother) home. The mother's daughter lives permanently with Family F but spends alternate weekends with her biological father (the mother's ex-partner). The children are exposed to different rules/guidance about alcohol when they are with their other parent/family. While the parents in Family F are aware of these differences and it has sometimes caused them concern, resulting in a discussion with one of the ex-partners, nonetheless it has not led to conflict within Family F, or between this family and the children's other parent(s)/family. Rather, the parents/carers recognised that while they might have different approaches from the children's other parents/carers about how to teach young people to drink sensibly, nonetheless their fundamental attitudes towards alcohol are the same and so they use these specific moments of disagreement about rules/boundaries as an opportunity to talk to the children about alcohol and alternative approaches to it:

Mother: … I think it was last year or the year before, Andrew [her husband's son] was going to a party and Clare [the boy's biological mother] bought him four alcopops to take … which I absolutely disagreed with … Because he was only 14 or 15.
Father: 15 … I think he was 15 and he was going to a party and she said 'Well I'd rather know what he's drinking than him just drinking anything there'. So I could kind of see her argument but it wouldn't have been my approach. So I

think in a way, they [his ex wife and her new partner] do have the same morals, the same way of looking at it but sometimes act on it … you know, their decisions are different from mine. But not necessarily the wrong decisions … I may disagree with what they're saying but I wouldn't disagree enough to think well that's the wrong decision, it's just a different way of doing it.

(Parents, Family F)

Parents/carers in the other case study families, although not encountering tensions with ex-partners, were nonetheless also aware that other people including their own friends and extended families as well as other adults that they came into contact with might also influence their children's attitudes to alcohol in both positive and negative ways. Like Family F above they also regarded other people's drinking practices – particularly excess drinking in public space – as providing a useful resource to enable them to talk to their children about alcohol in a specific rather than an abstract situation, and to offer guidance about consumption without needing to do so in a context where they were reprimanding their children about their own behaviour. Indeed, examples from beyond the family context were seen as particularly useful for providing advice about alcohol in this way because they were less 'value laden'. Where parents/carers are concerned about their children's persistent exposure to 'inappropriate' behaviour outside the framework of their control in the family home they subtly attempt to limit this by restricting their offspring's access to specific spaces. This is done directly, by not allowing their children to visit specific friend's homes at times when adults may be drinking; and indirectly, by trying to influence their children's friendships to steer them away from their peers who were from families which they perceived to model inappropriate practices ('the wrong crowd'). Here, parents/carers regard choice of school as a strategic way of controlling the environments within which their children mix.

Parents/carers were also ambivalent about taking children to pubs/bars when they were going to be drinking without food. Only 16.9 per cent of those surveyed agreed (slightly or strongly) that this would be an acceptable practice. Most of the parents/carers interviewed restricted their children's exposure to public drinking to meals at child-friendly establishments but were careful to avoid them experiencing more 'traditional' adult-oriented venues. In this way, parents/carers implicitly transmitted a construction of their own home-based drinking as safe or unproblematic in comparison to drinking in public space (this will be returned to in Chapter 3):

> Father: I mean it's just kind of like common sense really, I don't think we're doing anything bad by having a drink in the home or if people come round, other people having a drink. Or even if somebody gets quite drunk, it's normally very kind of happy drunk experience, it's not like something that she'd be horrified at. But equally you know, I wouldn't dream of going out to a pub … on a Friday night with all the kids.
>
> (Father, Family B)

While parents/carers model the benefits and pleasures of drinking, the majority are also careful to avoid demonstrating the negative consequences of drinking to excess. Of the respondents to the survey 90.4 per cent disagreed (slightly or strongly) that it is acceptable for parents/carers to be drunk when responsible for their children who are in bed and 87.8 per cent disagreed (slightly or strongly) that it is acceptable for parents/carers to be hung-over when responsible for children. The interviewees suggested that this is both because of concerns for children's safety and because of the painful realities of caring for small children with a hangover.

Nonetheless nearly a fifth (19.4 per cent) of the respondents to the survey agreed or slightly agreed that it is acceptable for children to stay up late at parties or events where parents/carers or other adults are drunk; an eighth (12.5 per cent) said their children had seen a parent with a bad hangover and 2.4 per cent stated their child had seen a parent sick or vomit because of excess alcohol consumption. Moreover, the qualitative case study research suggests that these survey results may underestimate children's exposure to drunkenness. First, because parents/carers rationalised that there are degrees of drunkenness and that while their children may have seen them behaving in 'silly' or 'happy' ways when they have been drinking this is still an extension of positive sociality of alcohol consumption, and is not the same as children witnessing the negative consequences of complete intoxication (such as falling over, being aggressive and so on). Second, because some of the children who took part in the case family interviews described occasions where they had seen one or both parents/carers drunk which their parents/carers had not defined as drunkenness, had forgotten about, or were unaware that the children had witnessed. Here, parents describe moments when their drunkenness had been observed by their children:

> Mother: I'm still embarrassed … that they actually saw me that drunk, I just think it's wrong … And it hit me and I realised just I was really drunk and I thought, oh the kids are up and they know and they've seen me, oh my God, this is awful and I was just like 'I'm really sorry that you've seen Mummy drunk'. And they were just like 'Yeah, yeah'. … I just thought … although they're aware of alcohol and we'll drink around the children but I don't like them to see that, I just don't think it's responsible. It wasn't responsible drinking and I don't want them to think that that's what you do.
>
> (Parents, Family B)

> My Dad is sometimes [drunk], they don't pick up on it, they don't realise that people are tipsy and they've had too much. They really don't. I mean we've had a couple of house parties in the evening and they've stayed up a little bit later and perhaps some of our friends have been a bit tipsy. I mean they haven't been falling over on the floor or anything like that, that the kids have seen … so I think they'll slowly be introduced to a bit of excess.
>
> (Parents, Family D)

As well as transmitting messages about when and where it is appropriate to drink through their own drinking behaviours parents/carers also intentionally model drinking practices to their children by socialising them into the rituals of drinking and encouraging them to try alcohol. Over two thirds (67.5 per cent) of the survey respondents reported that their eldest child in the age range 5–12 had been allowed a taste of alcohol, 23.1 per cent to sip an alcoholic drink and 2.6 per cent to have a watered down drink. The most common drinks tasted in this way were wine (56.6 per cent) and beer/lager/cider (52.4 per cent). Three quarters of the parents/carers who responded to the survey said that their eldest child in the age range 5–12 had tasted alcohol (with/without permission) at home (76.1 per cent). The case study research suggests that experimentation is commonly instigated by parents/carers rather than by the children themselves:

> Father: We have introduced it slightly by saying 'Do you want a bit of cider, with a bit of lemonade?' … And you know, even if I have a can, he doesn't

even ask … I will say 'Do you want some?' He doesn't come up and ask for it; it's me that asks him if he wants some and a lot of times he'll say 'No'. So it's kind of very hit and miss with him, he's not bothered at the moment. I think he's still very much a child and there's much more important things … he'd rather have a milkshake (laughs).

(Father, Family B)

In addition, to encouraging children to try different types of alcohol some of the case study families had also encouraged children to imitate adult drinking practices or rituals albeit with non-alcoholic drinks. These included giving children non-alcoholic cocktails that were modelled on the parents'/carer's own alcoholic drinks, teaching children how to drink shots of water, and asking children to fetch alcohol at home or to serve alcohol to visitors:

Father: I would drink beer in the house and I used to have Miranda trained, she used to … when she was little, I had her trained quite well where she would go to the fridge and get me a can of beer.

(Father, Family H)

Parents/carers also unintentionally use soft drinks to model the way that they themselves regard alcohol: as 'naughty but nice'. For example, in many of the case study families, sugary drinks such as colas, some types of fruit drinks and fizzy drinks were regarded by parents/carers as bad for children's health because they were perceived to rot children's teeth or to induce hyperactive behaviour. This construction has obvious parallels with the way that alcohol is perceived as a potential risk to adults' health and as inducing them, when drunk, to behave in abnormal ways. Just as most of the adults interviewed were aware of the risks of drinking above recommended limits and regard alcohol as a treat or reward, so too their children were being made aware of the potential health risks of particular soft drinks and were not allowed to drink these products regularly by their parents/carers – except as treats: at weekends, on special occasions (e.g. celebrations, parties, when eating out), on holiday or when they were being rewarded for good behaviour (Chapter 4 returns to such practices, from the viewpoint of children). In this way, a reverse morality of drinking is constructed within families where 'good' behaviour (by parent or child) is rewarded with a drink that could be potentially 'bad' (if health advice is disregarded) for the consumer (cf. James 1990).

Studies of the modelling of gambling behaviour within families suggests that it commonly follows gender lines: although having a father who is a problem gambler increases the risk that a son will follow suit, more than having a mother who is a problem gambler raises the likelihood that a daughter will do so (Walters 2001). Likewise, a study of family health and lifestyles found that mothers' health-risk lifestyles transmitted only to girls; whereas father's health-risk lifestyles transmitted only to boys which the researchers argued was a product of the fact that children and parents/carers are more likely to spend time interacting in same gender pairs (Wickarama et al. 1999). Yet, the evidence of the case study research is that drinking practices are more commonly modelled from father to daughter than father to son or mother to daughter:

> Father: Anne's had the odd sip of our drink, alcohol isn't the nicest tasting drink. It's not sweet. So you know, it's surprised me when Anne's had a sip and she's gone 'Oh I want some more'. Is it because obviously that I've got it, that she wants it? It surely can't be the taste.
>
> (Father, Family B)

This appears to stem from the fact that fathers tend to drink more alcohol at home than mothers and that girls in the age range 5 to 12 appear to be more interested in the lives of adults and therefore to imitate parental behaviour than boys, reflecting wider gendered patterns of social competence.

Conclusion

This chapter has focused on the transmission of attitudes and practices in everyday family life – an aspect of social reproduction that has been relatively neglected – through the lens of alcohol. In doing so, we considered parents'/carers' ideals of how family life should be – in Gillis' terms 'the families we live by'. In the case of alcohol, the dominant attitude of parents/carers surveyed and interviewed for this research, was that: children in general should not be introduced to alcohol at home until their mid teens; ought not to be allowed to drink in public space until they reach adulthood; and should be first introduced to alcohol by families at home (thus implicitly constructing an understanding of home as a safe space to drink in contrast to the risks that they associated with alcohol in public space). These were implicitly represented as an extra-familial 'norms' or lay moralities. In reflecting on their idealized notions of how children ought to be introduced to alcohol most of the respondents drew on stories about their own childhood family lives as a resource (to be repeated or reacted against). In this way, past family lives – even when negative or ambiguous – become reference points that help to shape present and future families.

We then went onto use case study research to explore the often taken for granted realities of *'the families we live with'*. Individual families create spaces and practices that are meaningful to them, albeit often taken for granted. In the context of this study on alcohol, these appear to be, in Fevre's (2000) terms, demoralized because they are not guided by the extra-familial norms or lay moralities parents/carers articulated, but rather are usually predicated on more pragmatic, common sense ways of thinking. Here, most of the parents/carers modelled a positive attitude towards alcohol – emphasising pleasure and sociality (notably a reverse morality that good behaviour can be rewarded by a 'naughty but nice' drink) – through their domestic drinking practices. These intra-familial 'norms' included encouraging children (aged 5 to 12) to try alcohol and to participate in drinking rituals (albeit often by mimicking these practices and moralities with soft drinks) at a much earlier age than government guidelines recommend and extra-familial norms suggest are appropriate. In this sense, extra-familial norms are somewhat of a moral façade because they do not guide or modify actual family practices. Despite the messiness, disorder, ambiguities and contradictions modelled by everyday family life, extra-familial moralities (i.e. about how people ought to behave) are nonetheless still transmitted to children. Yet, the unpredictable flow of daily activities and the inconsistencies of family behaviour are to-date not well accounted for in understandings of, and theorizing about, families.

The case study research also highlighted that most parents/carers adopted an individualised approach to parenting about alcohol. They were reluctant to reprimand other people's children for inappropriate drinking or for other adults to discipline their own children; resistant to the suggestion that alcohol education should be provided at school; and considered that general advice in relation to alcohol would be ineffective, believing that each child had an individual personality and needed to be parented in specific ways. In emphasising children's expressivity, rather than parental discipline, the interviewees presented families as resources out of which individual children construct themselves – in which children's identities are implicitly understood as fluid and self-determining. Parents/carers thus define the role of the family as to equip children with the right personal qualities and skills to ensure that they make 'sensible choices' in relation to alcohol. This neoliberal model of parenting/family life assumes that a child is able to distinguish between what might be the right action for himself/herself in a particular time and place. Yet, as some parents/carers observed when complaining about others who allow their offspring to binge drink in public space – without appreciating the significance of their comments – not all children have positive family support and as a consequence some are much less well equipped to make 'sensible choices' than others.

The emphasis parents/carers placed on choice does not however recognise that an individual's drinking can impact on many other lives beyond his/her own and consequently it does not acknowledge the wider shared social responsibilities of adults or children. This matters because the mundane practices of everyday family life are systematically linked not only to the well-being outcomes of individual children but also of society as a whole. The example of alcohol and family life thus illustrates a much wider process in which the rise of an ethic of self-interest is leading to a process of de-socialisation in which the public interest and importance of values such as social responsibility, trust and social cohesion are casualties. We argue that the importance of personal choice needs to be articulated within a broader framework of social obligation and relational morality. Consequently, the evidence of this research is that education – despite parents'/carers' resistance – remains an important way to address the gaps in what children are learning within families and the differential levels of education and support children receive at home.

Chapter 3
Rules and Guidance

This chapter considers the 'rules and guidance' bound up with transmission of drinking cultures within families. In particular we focus in more detail on the differential and discursive construction of the home as a space where parents/carers are happy to introduce children to alcohol in a 'safe' environment in opposition to public spaces that they consider to be locations where alcohol consumption is associated with violence and disorder (see Jayne et al. 2015). We argue that parents/carers miss opportunities to teach children about the range of drinking practices and spaces they may experience throughout their lives and fail to engage with their children about wider social responsibilities as potential drinkers in the future.

'It's All Part of Growing Up'

For all of the families who took part in our study the role of alcohol in parents'/carers' and children's current and future lives was an issue of significant concern. Nonetheless, despite its perceived importance, the differences between how attitudes and practices towards drinking in adults and children's lives are translated into rules and guidance were rarely simple and straightforward. As discussed in the previous chapter, patterns and contradictions that circulate around parents'/carers' attitudes towards alcohol consumption emerged from a reflection on the relationships between parents'/carers' own childhood experiences and their current ideals of how they *ought* to instil in their children a 'sensible' approach towards alcohol. For example, respondents were asked whether they thought that family drinking habits have changed across the generations. Thinking back to when they were 5–12 years old, and comparing that to today, 77.5 per cent of respondents stated that the range of alcoholic drinks parents/carers regularly enjoy has increased, 73.4

per cent of respondents stated that the amount that parents'/carers' drink at home compared to the pub has increased, 68.8 per cent of respondents stated that the amount that mothers drink has increased; and 65.7 per cent of respondents stated that the amount of alcohol that parents/carers drink in one session has increased.

These patterns reflect the fact that parents/carers believe that alcohol in the UK today is both more affordable (62.8 per cent) and increasingly marketed and advertised (63.3 per cent) than during the period of their own childhoods. These perceptions are borne out by broader research findings, for example, according to a review by Smith and Foxcroft (2009) alcohol is 65 per cent more affordable now than in 1980 and accounts for only 5.2 per cent of household spending compared to 7.5 per cent in 1980 (Office for National Statistic 2007). Most notably, the price of wine has fallen relative to average earnings (Mintel 2005). However, while research (e.g. Bogenschneider et al. 1998) suggests that that young people today are drinking at an earlier age condoned by parents/ carers than previous generations, the respondents to this survey were more ambivalent about whether children's access to alcohol has increased. Whilst 38.3 per cent of respondents thought that parents/carers are allowing their children to drink at an earlier age than during their own childhoods, nearly 27.1 per cent stated the age of permitted consumption was unchanged, and another 28.3 per cent thought this age had actually increased. Indeed, most of the parents/carers who were interviewed as part of the qualitative element of the research recalled being introduced to alcohol at home as children by their own parents/carers (e.g. trying wine at meals or Christmas/special events) and witnessing their parents/carers drinking at home and hosting dinners or parties where alcohol was consumed at a relatively early age.

Respondents nonetheless expressed high levels of concern over perceived shifts in both public drinking cultures and the nature of young people's drinking since their own childhoods rather than worrying changes in 'private' or intra-familial drinking cultures. These views correlate with intergenerational studies of continuities and change in UK drinking cultures (Valentine et al. 2010b). Notably, several of the interviewees claimed that when they were young it was less common to see either adults or young people drunk and 'behaving badly' in public space; and that their own underage drinking had been motivated by sociability and a desire to have fun, not by the aim of getting drunk (a contemporary practice that Measham 2006 has labelled 'determined drunkenness'). While alcohol was a part of their teenage social lives, most of the interviewees nonetheless had few recollections of peer pressure to drink to excess until they were young adults at college/university or in paid employment.

As such, some of the interviewees made comments about how children are becoming older younger which accords with Postman's (1982) thesis about the 'disappearance of childhood':

> Mother: We didn't drink loads … if we had a full night out, we'd perhaps get two Martinis (laughs), with lemonade … We [she and her then boyfriend, now husband] used to be in the Venture Scouts and we used to go on holidays and we'd get a bit merry but not drunk-drunk … We're more likely to go out and get drunk now than we were back then … If we go out now, we usually have quite a few … I mean even when I was 19/20, going out with some old school friends, nobody got drunk …
>
> Interviewer: Right. I mean why do you think people didn't?
>
> Mother: It wasn't the culture, back in the 1970s/80s, people didn't, you didn't have this culture where you'd got into town on a Saturday night now and people are staggering, you didn't get that. People would have had a drink and been quite merry and whatever but you didn't see people being sick or getting … it just didn't happen … It wasn't as intimidating going out. I mean I wouldn't dream of going to town on a Saturday night now.
>
> (Mother, Family C)

> I think just from the general kind of stuff that you listen to on the news and in newspapers and magazine, it just all seems to be happening a lot, lot earlier … I just think there's a natural instinct to experiment at a certain age and I think that is lower nowadays [edit] … I mean at 13 I started thinking about a lot of things but maybe didn't actually action things until a bit later, about 15. And that had nothing to do with what my parents had instilled in me.
>
> (Mother, Family E)

Indeed, only a relatively small percentage of parents/carers who took part in our survey believed that their own children (aged 5 to 12) had actually drunk alcohol either in their own or other people's company; 25.0 per cent of respondents said children had tasted/drunk alcohol at a family celebration, 10.0 per cent said their children had tasted/drunk alcohol at a relatives/ friends home, 3.0 per cent said their children had tasted/drunk alcohol at a pub restaurant, 1.0 per cent said their children had tasted/drunk alcohol at a public event, and 0.2 per cent said their children had tasted/drunk alcohol with their children's friends outside their home. However, in contrast to the questionnaire survey the qualitative case study research – which included families with very diverse parenting styles and different levels of personal alcohol consumption – found that the majority of their children (aged 5 to 12) had either been offered or had tried alcohol at home, with a family meal or at a family event. These domestic practices actually accord closely with the parental interviewees' recollections of their own childhood experiences of being introduced to alcohol.

The contradictions between the quantitative and qualitative elements of the research can be explained in one of two ways. Firstly, the under reporting of alcohol consumption in questionnaire surveys is a widely accepted phenomenon. Indeed, there was a discrepancy in our survey between parents'/carers' perceived and actual levels of consumption. This raises the question of whether the parents/carers surveyed may have under-reported their knowledge of their own children's consumption of alcohol because of the 'sensitivity' of this issue and a desire to fit in with perceived extra-familial parenting 'norms'. Second, parents/carers may have responded to the survey with their ideals or how they intend/think they ought to parent which while being a genuine response may not be how they actually parent in practice because of the pressures of everyday family life that mean that intra-familial lived realities do not always accord with extra-familial ideals (cf. Valentine 1999). With contemporary discourses regarding parenting circulating around 'closer' and less hierarchical relationships between parents/carers and children (Valentine 2004), what is clear in the results of our study is that there has been a significant generational shift in parents'/carers' attitudes to alcohol consumption within the family.

Moreover, in contrast to their own childhoods parents/carers sought to develop close and loving relationships with their children by establishing practices of openness and good communication within their families that included directness about alcohol (see Valentine et al. 2011). This pattern of a desire for open relationships was also evident in the survey responses: where only 14.7 per cent of the respondents agreed strongly with the statement 'I try to protect my child, and don't talk to them about adult topics':

They were a very strong influence in how I feel about things. But I think that's with everybody you know, you draw on your own experiences and want to better them. And also I think oh you know, you think back and you think oh that would have a real negative effect on me, I'd never do that with my children, or I would do things differently. Do you know what I mean, you make your own judgments from your experiences I think. And also good things as well, I mean certainly a lot of the good things I think what my Mum did with us you know, I've done with my children [edit] Yeah, the bad thing, she was extremely protective because I think she knew she had to be the sole parent she … wouldn't let us play out very much with our friends or only for a short period of time. When we got to secondary school, when there were school discos, we weren't allowed ever to stay to the end, she always insisted on picking us up half an hour before the end … So she was overprotective and because of that we retaliated and we were hard work and in trouble a hell of a lot.

(Mother, Family D)

I want to be more relaxed with my children. Because I think … I mean there's loads of things I didn't tell my Mum as I was growing up, loads of things which I wished I could, and I wouldn't want my children to feel that way, which is why I think I want to be more open and more relaxed … I want them to be honest. That's the one important thing to me … I think in time that kind of makes the relationship closer, makes them feel a bit more secure. And if anything you know, upsets them at school or friendships when they get older, then hopefully they'll be happy to talk to me about it and hopefully we'll sort it out.

(Mother, Family E)

However, parents/carers intentions to be positive and open with children was not straightforwardly translated to specific rules and guidance. In the next section of the chapter we therefore focus in more detail on gendered geographies of parental concerns regarding potential threats to children regarding alcohol consumption (Valentine 2004; Wickama et al. 1999). In particular we show that it is important to understand the transmission of drinking cultures through consideration of 'what it is to be a child' is bound up with interactions with adults in particular ways, in specific spaces and places. For example, evidence from Europe, Australia and New Zealand highlights how drinking practices of children/young people come into conflict with adults in certain spaces in residential neighbourhoods, parks, beaches, and so on with specific reference to noise, exuberant behaviour and specific types of alcoholic drinks (Popham 2005; Jones 2002; Kelly and Kowalyszyn 2002; Kraack and Kenway 2002; Mackintosh et al. 1887). Indeed, from such writing and our own research findings it is clear that issues bound up with intergenerationality are central to understanding the conflicts between intra and extra-familial influences of the geographies of the transmission of drinking cultures.

Who is Most at Risk and Where?

Mike Leyshon (2008b) highlights that discourses around childhood/young people are imagined, defined and created around certain place myths and practices. However, Leyshon also considers the social and material relations predicated on how young people actively produce understanding, belonging and not-belonging and the sometimes contradictory feelings of inclusion and exclusion in ways that can be very different from adults. Leyshon (2005, 2008a) thus describes the performing of moral geographies and imagined deviance relating to young people and drinking in public space in terms of a process of problematization and reconstruction. Thus, while most of the parents/carers interviewed did not feel their children aged 5 to 12 faced any current alcohol-related harms, many of them expressed concerned attitudes about the potential risks their children may face as teenagers as they become more 'independent' and are exposed to and experience alcohol consumption beyond the home. These risks of future dangers were perceived to be highly gendered: with girls described as being vulnerable to sexual violence when drinking in public space, and boys as vulnerable to getting involved in fights or other kinds of social disorder in the night-time economy (see Holloway et al. 2010; Jayne et al. 2015).

These views mirror wider social attitudes about gender and alcohol consumption. However, while there has been a significant increase in drinking by women with the gender gap between men and women's drinking behaviours narrowing in the last 30 years, moral attitudes towards women's drinking in public space have not kept pace with this social change (Smith and Foxcroft 2009). Women drinkers still face more opprobrium than their male counterparts, reflecting the persistence of traditional gendered (and classed) expectations of 'respectability' and historical sexual discourses about women in public space as 'loose', and as inviting male violence (Plant 1997; Day et al. 2004; Ettore 1997):

> Father: I think they're natural worries that everyone has. But I think with the girls, I know it sounds sexist but I always feel like girls need more protection. I think as teenagers, they are almost more vulnerable ...
> Mother: I think ... well I'm speaking from personal experience where sort of you think you are indestructible and it'll never happen to me, and I look back now at things that I did and think I don't want my children to do that. I don't want them going out wearing high heels and short skirt and then walk home on their own, thinking 'Oh nothing will happen to me', when clearly it could.
> Interviewer: So the risks are slightly different?
> Father: The risks are different ... I just think there is a responsibility that both boys and girls need to understand, it's okay to go out and get drunk but don't go out and get so drunk that you end up doing silly things or getting yourself into trouble ... And I think that applies to both ...
> Mother: And do it in a safe environment, if possible, which would be at a house party.
>
> (Parents, Family F)

> No, I mean obviously I don't want them to be raving drunks or whatever and I suppose I'd be more concerned about my daughter getting drunk at parties and getting into trouble ... Then the other side is lads going out and getting drunk in town centres are at risk as well, so.
>
> (Father, Family D)

Bound up in these interview quotes, is an implicit attitude that some spaces are 'safer' than others (see Holloway et al. 2008). Here, young people's drinking is considered more risky outside the home because the extra-familial norms associated with young people drinking in public space are predicated on representations of 'bingeing' and 'anti-social' or irresponsible behaviour. In these terms, parents/carers generally expressed a notion of their own childhoods as being 'safer' and their own behaviour as more 'sensible' and in doing so clearly 'bought-into' popular and policy discourses relating to 'moral panics' of young people being drunk, violent and disorderly with little attempt to entertain a more balanced understanding of the pleasures and dangers of alcohol consumption for contemporary children/young people (see Jayne et al. 2015).

Indeed, as noted in Chapter 2 some of the interviewees placed responsibility for young people's problem drinking in public space on what they perceived as poor parenting by 'others':

> It's better than on street corners. Yeah, I said 'I'd rather you [to her older daughter] were sat in [name of local pub removed] with a group of friends responsibly'. And I did point out to her that she could go in [name of pub] and drink soft drinks at 16 and it's a social thing for her every week now [edit] … And I think well it's up the road, we know where she is and you see so many kids on street corners with bottles of Vodka and cans of cider.
>
> (Parents, Family C)

> I'd rather do it that way [allowing his son and friends to drink in his home], then I know what he's doing and then hope that by doing that, it will open a link of trust between us both … once they're out of that front door, you're not going to know until you get the knock on the door from the Police or whatever, saying 'He's been found drunk'. You know, I'd be quite happy to supervise him …
>
> (Father, Family B)

> Mother: But I don't think binge drinking in itself is the problem, I think it's the type of people who do the binge drink [edit] … It may be I think you know, how they were brought up. I just think however drunk you are, your basic morals will still be there. So yes, I might get really drunk and steal a traffic cone, which is you know, when you're at university quite funny, but I can't imagine … that I would ever get really drunk and smash a bottle in someone's face. You just know … however drunk you are, you have your limit of what's acceptable and what really isn't, what's unacceptable social behaviour.
>
> (Parents, Family F)

However, contrary to a common perception that 'other' people's parenting is at the root of young people's alcohol-related anti-social behaviour most of the interviewees argued that home was the main site where children should learn about alcohol rather than school. While some interviewees noted the potential role of education in reinforcing messages about responsible drinking and meeting the needs of those who do not get support at home, most parents/ carers felt that their children needed tailored advice and guidance which they were best placed to offer because of their intimate knowledge of their children, their friends and the spaces that they inhabited and consequently the people and situations they were likely to encounter:

> Father: Definitely from the parents. I think that should be the first introduction. I don't think schools should really …
>
> Mother: It's not appropriate really in schools.
>
> Father: No, schools are to educate you, not to learn about drinking.
>
> Mother: Yeah, it's not something … I think introduction definitely within the home life.
>
> Father: Not to set up a whole lesson in school in the curriculum on drinking or whatever. I mean there's talk about changing school curriculums to include all these sort of life skills, but I mean that's almost taking over what parents should be doing …
>
> Mother: You're right but a lot of parents don't and I think that's why they're looking to introduce it or do more in school because a lot of kids don't have as much guidance as you or I might be able to give them …
>
> Father: It's a vicious circle because the parents don't put enough effort in because they know the school's going to do it.
>
> (Parents, Family D)

Such evidence highlights how 'rules and guidance' relating to 'problematic', 'sensible', 'safe' and 'unsafe' drinking practices, and the spaces and places where alcohol consumption takes place are differentially and discursively constructed in relation to each other. The work of Evans (2008, 2010) is particularly important for understanding the pre-emptive logics and politics of restrictions put on young people's use of public space. Evans (2008: 1662) considers the spatialities and boundaries between children/young people and adults as being far from fixed and for the need to consider the 'changing social constructions of childhood and adolescence and the processes that structure young people's lives across a range of spatial scales'. Moreover, advancing debates about relationality and intergenerationality, Evans (2008: 1669) shows how recent work on movement to 'independence' has focused on 'developing a more nuanced understanding of the negotiation and ongoing *inter*dependence with significant (extra)familial others during key 'transition events [and that] … 'dependency' and 'interdependency' having moral implications for both young people and parents'. Such insights are important in understanding the complex adult/young people/children relationships bound up with geographies of alcohol, drinking and drunkenness.

The ideological dimensions that constitute notions of, for example, the home and associated geographies of alcohol consumption were thus shown to relate to visions of urban drinking practices and experiences relating to violence and disorder and the role and influence of other adults/children/young people. Moreover, the findings show how links between home and public spaces relate not only to alcohol consumption per se, but are reflective of concerns regarding the 'wider world', which as parents/carers in the study highlight are not necessarily informed through personal experiences, or indeed the reality for the vast majority of adults and children/young people who drink alcohol in public spaces but via geographical imaginations bound up with wider extra-familial political and popular debates and representations of alcohol, drinking and drunkenness.

Conclusion

In this chapter we have seen that parents/carers perceive that there has been a significant shift in public drinking cultures and the nature of young people's drinking since their own childhoods. Such understanding of geographies of alcohol, drinking and drunkenness highlight 'problematic', 'sensible', 'safe' and 'unsafe' drinking practices bound up with specific spaces and places, differentially and discursively constructed in relation to each other. Despite these insights there is still significant theoretical and empirical work that needs to be done in order to draw out the connections and mobilities that constitute adult geographies of and the diverse range of spaces where children encounter alcohol in everyday settings. Sustained engagement with such relationalities and the complex intergenerational conflicts, tensions and dialogues that are bound up with adult/young people/children's geographies can thus be seen to offer much to advancing understanding of the transmission of drinking cultures and 'rules and guidance' within families in a way that also highlights the false dualisms and mis-conceptions associated with 'public' and 'private' drinking cultures.

Chapter 4
'Drinking is for Grown Ups'

This chapter focuses on children's knowledge and understanding of alcohol, drinking and drunkenness addressing three questions: what do pre-teen children know about alcohol and how do they learn about alcohol and its associated potential harms? We show that while parents/carers are ambivalent about talking to pre-teen children about drinking, regarding them as too young for such discussions, the children themselves have developed a competent understanding of alcohol and the circumstances under which children and adults may drink. Much of this knowledge about alcohol is gleaned by children through proximal processes namely their daily interactions with parents/carers/older siblings in the context of everyday family life that are unintentionally modelled to children.

'When You are Big You Can Have a Whole Bottle': What Children Know About Alcohol and How They Learn About It

Safe, Sensible, Social, the second phase of the alcohol harm reduction strategy for England and Wales (Department of Health and Home Office 2007) identified the important role of parents/carers to provide young people with information about alcohol, and to support them to make responsible decisions about its consumption. Yet, most of the parents/carers interviewed were reluctant to address this issue with children considering them too young to understand complex health messages about such an 'adult' topic. However, while drinking is commonly associated with teenage years, previous research has shown that from the age of 6 children understand the concept of alcohol (Jahoda and Cramond 1972; Casswell et al. 1988). This was evidenced in our case study families. All of the children (aged 5–12) interviewed understood that alcohol is an adult product, although they had only a sketchy understanding of the legal framework relating to children's alcohol consumption (e.g. the age at which children can enter a pub or buy a drink). This distancing of childhood from alcohol is evident in the following quotations:

> Interviewer: … when can children start to have these drinks?
> Girl: Well children can't have them but when they grow up they can have them.
> Interviewer: So how old do they have to be?
> Girl: 36 or maybe like 49 or something like that.
>
> (Lucy, aged 7, Family I)

> Interviewer: Can children go in pubs?
> Boy: Yeah, sometimes.
> Interviewer: Could you go into a pub on your own?
> Boy: No … because I'm not over 13 years old.
>
> (James, aged 9, Family D)

The reasons children gave for why only adults are permitted to drink alcohol hinted at an awareness of some of the embodied consequences of its consumption including recognition that: alcohol will affect children more rapidly than adults; it has both physical effects on the body and social effects on behaviour; and that there is a risk of addiction. For examples: 'Kids can get drunk quicker' (Karl, aged 11, Family B); 'Kids lose control more quickly' (James, aged 9, Family D); and [children] may not be able to stop' (Linda, aged 6, Family A).

Yet, the product recognition methods (by smell and advertisement) identified that the children had diverse, and in most cases an inaccurate grasp, of the alcohol content of different types of product and the number of drinks necessary to become drunk. A greater association was made between drunkenness and beer than wine which was perceived to take longer to consume and weaker. This misunderstanding about relative alcohol content may stem from the influence of television advertising where wine is commonly gendered as feminine and represented in terms of 'middle-class' practices of dining, whereas beer has more masculinist associations with pubs, sport, and violence. Some of the children also had a misperception about alcopops assuming these were mainly 'pop' (colloquial term for non-alcoholic drinks).

Jahoda and Cramond's (1972) study in Scotland found that young children were familiar with the names of some alcohol products: with those aged 6 able to identify at least one drink by smell, those aged 8 able to determine which from a selection of bottles contained alcohol. These findings have been replicated in subsequent studies elsewhere in the UK and US which have demonstrated that contemporary children aged 5 can recognise alcohol from pictures and have expectations about the role of alcohol in adults' social lives (Andrews et al. 2003). However, the children in our study demonstrated limited knowledge of specific types/brands of drink. While a few children did recognise specific drinks from films and television programmes (the farmer drinks cider in the film *Fantastic Mr Fox*) and stories (gin is referred to in *George's Marvellous Medicine*) the majority *only* correctly identified the alcohol that their own parents/carers or relatives drank (including older siblings and grandparents), including in some cases recognising gendered product preferences and the consumption of different drinks (e.g. cocktails and shots). These patterns highlight the significance of 'modelling' in everyday family life as a process through which attitudes and practices are transmitted. The children's interviews (using a doll's house game and puppets) also demonstrated how they had picked up the association of alcohol with friendship and sociality which is modelled through parents'/carers' practices. Children from five of the case study families described having seen a parent or sibling drunk although the occasions which they described were commonly related to parties or holidays, reflecting the observation (above) that parents/carers are often unaware of the significance of moments when they model 'abnormal' patterns of consumption:

> Girl: Beer [identifying a sample of alcohol by its smell].
> Interviewer: Beer, okay. So where have you seen that one before?
> Girl: Everywhere … Everywhere in the house, apart from upstairs.
> Interviewer: And who drinks that one?
> Girl: Daddy.
> Interviewer: Have you tasted it ever? Do you know what it tastes like?
> Girl: Yeah, I do like it but Daddy doesn't let me have a lot … because it's too fizzy.
>
> [edit later]
>
> Interviewer: [Showing a picture of a brand of beer] And who drinks that one?
> Girl: Daddy and me.
> Interviewer: And you? How many times have you tasted beer do you think?
> Girl: Five or six.
>
> (Mel, aged 5, Family D)

These patterns highlight the potential significance of proximity in children's development of knowledge about alcohol. Namely, intimate embodied practices, which mediate relationships between parents/carers and offspring in the interiority of family life, have the potential to advance children's individual learning about alcohol beyond that which they might gain independently from observing the public realm:

> [Identifying a picture] It's beer and it's called John Smiths ...
> Interviewer: Who drinks that one?
> My Daddy. Sometimes my Daddy drinks it [... discussion of other pictures]
> Smirnoff, my Mummy's favourite ...
> Interviewer: Does anybody else drink that one?
> Auntie Nina.
> Interviewer: ... And have you tasted that one?
> No, I'm not allowed it! [laughs] ... It's my Mummy's favourite but she never lets me have it. Even she's got her own Smirnoff glass.
>
> (Anne, aged 7, Family B)

> Girl: [identifying a picture of a drink] I've seen this before, sometimes Daddy drinks this one ... He drinks every single beer. Sometimes he might have some beer in the fridge.
> Interviewer: So it's in there [indicating the picture and the fridge]?
> Girl: Quite a lot of bottles ...
> [edit: discussion of other images]
> Girl: Guinness! My daddy drinks that at Christmas.
>
> (Miranda, aged 7, Family H)

The children's interviews and games in which they were invited to use a dolls house and puppets to show the researcher when they remembered seeing alcohol at home demonstrated that they had picked up a specific association of alcohol with friendship and sociality modelled through parents'/carers practices (although this is not to suggest that as adults they will necessarily use alcohol in the same way). Most of the children used the puppets to act out offering drinks to family members and visitors, pouring drinks for others, and having a party. Here, the children's identification with their parents'/carers' consumption of alcohol related to the positive emotional context of its use. This finding is counter to previous studies (e.g. Casswell et al. 1988) which have suggested that between the ages of 6–10 children generally have negative attitudes towards alcohol which may be explained by the contemporary increase in domestic consumption and the normalisation of alcohol within the home (Ogilvie et al. 2005; Holloway et al. 2008; Smith and Foxcroft 2009).

When asked about their probable attitudes towards alcohol and drinking practices when they are adults, the children interviewed anticipated a future of moderation. In particular, their imagined futures hinted at a sophisticated recognition that drinking alcohol is a pleasurable and social activity, while also showing awareness of some of the social risks associated with excess consumption (particularly drunkenness), despite their generally limited or confused understanding of the possibilities of alcohol-related harms to physical health. In this sense, their views generally replicated the balanced attitude towards alcohol that their parents/carers suggested that they wished to instill in them:

> Interviewer: Do you think you'll drink when you get older?
> Boy: Probably, yeah … maybe beer or something … I don't think I'd ever get drunk.
> Interviewer: Right, so you think getting drunk is bad?
> Boy: Yeah … Because it makes you go a bit crazy in the head, like I said before.
> Interviewer: … does it have any other effects on your health? …
> Boy: I think it like damages your lungs or your heart or something.
>
> (John, aged 9, Family E)

A comparison of the interviews with parents/carers and children within each case study family identified the transmission of future consumption intentions within specific individual families again predicated on the positive emotional context of its consumption. In this sense children's affective ties with parents/carers appear to intensify their learning about alcohol. For example, Karl (aged 11, Family B) said that if going to a party he would take Smirnoff – which he had previously described as his mother's favourite drink and which her friends bring when they visit. Likewise, Family A enjoy whiskey and describe themselves as 'connoisseurs'. This narrative of distinction, which was evident in the parents'/carers' interviews, was echoed in the children's accounts where the girls' referred to 'precious drinks' and associated alcohol consumption with the Queen and being 'posh': a construction of drinking practices that was not evident in other children's narratives.

The case study research also shows that experimentation is commonly instigated by parents/carers not children and that families also encouraged children to imitate adult drinking rituals albeit with non-alcoholic drinks (this will be returned to later). In these ways, parents/carers advance children's knowledge about alcohol within the interiority of family life:

> [Identifying a picture] Cocktails! … Sometimes my mummy … makes a special one for me … mummy's has got alcohol in it. I don't have alcohol in mine.
>
> (Anne, aged 7, Family B)

> [Identifying a picture] Alcohol, it's a shot of alcohol.
> Interviewer: And why do you only get that much, do you know?
> … because it's a very strong alcohol … You just go … like that [imitating knocking back a shot]. I've actually drank … I've actually put some water in there and … my dad asks me to see how long it takes me to drink four of them.
>
> (Karl, aged 11, Family B)

When asked to name places which they associated with alcohol children's most common response was to name a supermarket rather than traditional venues such as the pub, bar or the off-licence. A few of the younger children named atypical locations where they had seen a parent drink ('at my school summer fayre when Mummy was having it' – Lisa, aged 6, Family A) further demonstrating the significance of proximity and the socially embedded nature of children's learning about alcohol. Children were also aware of the age identification campaign (Challenge 21) necessary to purchase alcohol in supermarkets, raising this unprompted, although some were confused about the age at which it is legal to buy alcohol:

> [Referring to a picture] Well that one would be for grown-ups only because it's alcohol.
> Interviewer: … so how do you know that one's got alcohol in?
> I just do … If you go in and you have a look, it would say this is for grown-ups only.
> Interviewer: Would it?
> Yeah, or over 25 because in Tesco, if you go into the wine column, it says if you look like you're under 25 … they have to look at your driver's licence [edit]
> Interviewer: Tescos, are there any other places where you can buy those drinks?

You can probably get some in Waitrose, you can get some in Marks & Spencer's.
(Aileen, aged 8, Family G)

Girl: Quite a lot of the shops but when you go into a shop, people under like about 21 or something, they say … I think they say 'How old are you?' and then you have to say how old you are. And then if you're like 21 or older you can have it but if you're under 21, you can't.
(Gemma, aged 9, Family A)

Girl: Because well … I don't know. I know sometimes if you crash or things like that, they check if you've like had alcohol and things. And if you do … if you have had alcohol, you could get your licence taken off you I think.
Interviewer: Right, so there's a law about that?
(Linda, aged 6, Family A)

This general association by children of alcohol with family shopping routines reflects the changing geography of alcohol consumption – from a predominantly public practice in specialist locations (such as the pub), to an everyday domestic practice facilitated by the increased affordability of alcohol (its real price has halved since the 1960s), and the growth of off-trade sales (e.g. via supermarkets) (Ogilvie et al. 2005). This growth in home-centred drinking (Holloway et al. 2008) means that children's indirect proximity to alcohol from observing everyday familial behaviours, as well as direct access to alcohol at home, has increased significantly. This proximity effect highlights the importance of the experiential (e.g. habitual routines and shared family practices) in terms of the way that children develop knowledge about alcohol.

Zig-zagging Around and Going Crazy in the Head: What Children Know About the Harms Associated with Alcohol and How They Learn About Risk

Most of the children had a good general understanding of what a drunken person looks like and how they behave from popular culture ('look tired', 'eyes half closed', 'smell of drink', 'walk strangely', 'zig-zag about'). Their observations tended to focus on the short-term behavioural or social effects of alcohol. They also repeated public-education messages about drink-driving which they had learned from television campaigns, with several children nuancing these warnings with specific limits on consumption (it's permissible to have one alcoholic drink and drive) that they had picked up from their parents/carers. Albeit, some of the younger children were less clear about whether the restriction on drinking and driving applied only to alcohol or also to other 'adult' drinks like coffee. In this sense, the children had largely understood distancing messages about their separation from this adult world through the negative emotional contexts in which alcohol was presented (e.g. accidents).

Approximately 1 in 5 (20.6 per cent) of the parents/carers who responded to the survey said their child had ever expressed a concern about somebody's drinking: their own, their spouse/partner, ex-spouse/ex-partner, sibling, or a friend/relative. Approximately, 17.8 per cent of the respondents said their child had mentioned one individual as a source of concern, while 2.8 per cent said their child had concerns about two or more people. Children from five of the case study families described having seen a parent or sibling drunk although the occasions which they described were commonly related to parties or holidays, reflecting the fact that parents/carers are often unaware of the significance of such intimate familial moments when they model 'abnormal' patterns of consumption:

> Interviewer: Have you seen people drunk?
> Girl: No. Oh yeah, my Mummy [laughs] …
> Interviewer: And how can you tell?
> Girl: Because she throws up sometimes.
>
> (Anne, aged 7, Family B)

> Boy: my Dad once drunk alcohol but then he had to go to bed [edit]
> Interviewer: Do you think that you will have drinks when you grow up?
> Boy: No.
> Interviewer: No? Why not?
> Boy: No … Well I think when I grow up, I think my Dad might drink some alcohol and then he might fall asleep, so that's why I won't drink it.
>
> (James, aged 9, Family D)

Girl: When we went to Greece my sister, she had about one or two cocktails and when we went back to our … apartment and she went into her bedroom and she just laid down on the bed laughing …

Interviewer: So what does alcohol do to you when you drink it? Does it change people in any way?

Boy: It makes them a bit less controlled of theirself.

Girl: Yeah, my friend, she asked her Mum if she could have a dog … when she was drunk, and she went 'Yes, maybe'. And then she asked her the next morning [when the mother was sober] and … she was like she can't have one …

Interviewer: Yeah, it makes people say things that they don't maybe mean?

Girl: Well they sing stupid songs … my Mum and my friend's Mum got drunk … They was a bit drunk and they started singing a song about what you do when you need the toilet when you're working in the garden.

Interviewer: Right, so what's a bit drunk then?

Boy: Sort of a bit strange, a bit weird. Yeah, just a bit strange, not theirself …

Girl: A bit … like really messy hair.

Boy: She had hair all over the place before didn't she?

Interviewer: Did you think it was funny at the time … ?

Boy: Not a lot, no … Because it's not good for her.

(Emma and Tim, aged 10, Family C)

As noted in Chapter 2 children did not however appear to feel threatened or upset by adults' drunkenness. Rather, they commonly represented their parents'/carers' behaviour in a rather bemused way, although one child recognised that there are degrees of drunkenness and that if someone is 'a bit drunk' you can have fun with them but if they are 'very drunk' you should stay away. While the parents/carers were concerned that their children might be judgmental about their drunkenness, their offspring did not associate alcohol with moral failings, perhaps reflecting the extent to which domestic drinking, even to excess, has become normalised in UK culture.

However, when children talked about being drunk in abstract terms rather than in relation to family members they represented it in negative imagery, drawing a striking association between alcohol and aggression. Here, the negative portrayal of alcohol and violence on television were key reference points for the children's observations, as well as some recollections of seeing drunken strangers behaving in threatening ways in the street. Although television appears to provide an important source of risk information for children about the potential social harms of excess alcohol consumption in public, as shown in Chapter 3 the spatiality of their moral distinction between drunkenness at home and in public space suggests a problematic disassociation between children's understandings of the negative effects of drinking to excess and everyday family practices:

> Because alcohol has this sort of thing in that can make like kids do things that they're not supposed to do … like fight people and kill people … and kids aren't supposed to do things like that. And the other reason is it can damage them.
>
> Interviewer: How can it damage them, do you know?
>
> Because if you drink some, it can damage them because they might not be able to be like a proper person anymore [edit later] because if they have alcohol, it can make you really like naughty … They might punch people, sometimes say things that they don't mean, like 'I hate you' … if they have a lot they do nasty things but if they don't have that much, they're nice.
>
> (Linda, aged 6, Family A)

> Interviewer: … Have you ever seen anyone you know drunk?
>
> No. … I've only seen one or two [drunk strangers] and it's been after a football match. But one man, I was walking home from school and my Mum actually called the Police on him [edit]
>
> Interviewer: So do you think getting drunk is something that lots of people do or just a few people?
>
> Teenagers do it quite a lot … And they usually talk about murdering and things.
>
> Interviewer: Murdering?
>
> Yeah, other people and on the news and things they're just talking about teenagers, blah-blah-blah … they just do bad things?
>
> (John, aged 9, Family E)

While children demonstrated generally competent understanding of some of the potential social harms associated with excess alcohol consumption in public space they had a limited understanding of the long-term health risks associated with drinking above recommended limits. These are defined by the UK Department of Health as cancer of the mouth and throat, sexual and mental health problems, liver cirrhosis and heart disease for adults, as well as affecting brain development and the risk of accidents, injury and alcohol poisoning for children. Although previous research (e.g. Kurtz 1999) has suggested that health is not an issue of significant concern to children, perhaps because these risks are commonly framed in the future (e.g. Valentine et al. 2010), nonetheless those who took part in our study had a reasonable knowledge of the health harms associated with other social practices like smoking. Indeed, in some cases the children muddled the health warnings associated with smoking, drugs and alcohol. John (aged 9, Family E), for example, suggested that drinking might damage your lungs, Aileen (aged 8, Family G) thought that Michael Jackson had died from alcohol rather than drug consumption, while Lucy (aged 7, Family I) made a loose association that alcohol is more of a threat to children's health than adults' and can result in a heart attack. Where children were aware of the concept of addiction they associated it with warnings about playing computer games – the framework within which many parents/carers introduced them to this concept – rather than with alcohol:

> Getting addicted, like you've tried it and then you want to do it again and again and again … You can get addicted to a game, like Club Penguin …
> Interviewer: Do you know of any famous people who drink a lot of alcohol or celebrities?
> I know someone who did but he died.
> Interviewer: Who's that?
> Michael Jackson.
> Interviewer: Michael Jackson; he was addicted to alcohol was he?
> He died from it.
>
> (Aileen, aged 8, Family G)

The case study parents/carers were commonly ambivalent about talking to pre-teen children about alcohol, arguing that they lack sufficient understanding to receive such complex health information but nonetheless (as noted in previous chapters) described using soft drinks to model the way that they regard alcohol: as 'naughty but nice'. In most of the families, sugary drinks were considered bad for children's health and liable to cause hyperactive behaviour. This representation has parallels with the way that alcohol is perceived by adults as a potential health harm and cause of anti-social behaviour. In the same way, they warn their children about the potential health risks of particular soft drinks and do not permit them to drink these products regularly – except as treats or when they are being rewarded for good behaviour. In this way, a reverse morality of drinking is constructed within families (cf. James 1990 study of confectionary) where 'good' behaviour by a parent or child is rewarded with a drink that could be potentially 'bad' if health advice is disregarded by the consumer:

> Girl: It smells like lemonade.
> Interviewer: Okay, so you like that smell and you've had that before, that drink?
> Girl: Yeah because my brother, Neil, always has it … We're only allowed fizzy drinks in the holidays and things like that.
> Interviewer: Okay, why is that?
> Girl: It's because sometimes they have sugar in and it kind of makes you a bit hyper because Neil used to have it when … he tried it and he got a bit hyper … you can be hyper with them [both laugh] … we're allowed it at parties as well.
>
> (Aileen, aged 8, Family G)

> Interviewer: … if you could have a choice, what would you ask for do you think?
> Girl: Lemonade … I'm not always allowed it.
> Interviewer: Why not?
> Girl: Because Mum says it's really bad for my teeth … because it's sugary.
> Interviewer: Okay, so you don't have fizzy drinks very often? … What sort of occasions might you have them?
> Girl: If I be good.
>
> (Anastasia, aged 4, Family I)

However, while previous research has indicated that the place where children are most likely to obtain and consume alcohol is at home or their friends' homes, supplied by parents/carers, and despite the fact that all the children (aged 5–12) in the case study families were exposed to alcohol consumption at home, the majority had little interest in experimenting with it (e.g. Valentine et al. 2007; Hibell et al. 2009). Some had tried it but most either actively disliked the taste of alcohol or preferred soft drinks. This was further borne out by the evidence of participant observation at family events. Here, children showed little interest in alcohol despite the fact that adults' were drinking. Rather, the children commonly carved out their own 'private' space where they could play together independent from the adults' activities and were happy to enjoy their own 'treats' such as fizzy drinks without showing interest in what adults were consuming:

> Girl: He [her father] sometimes takes me to a pub, there's nothing to play with but you can run around a lot. [edit] I have mango juice and ham crisps.
> Interviewer: And are there other children there at the pub?
> Girl: Sometimes [edit] There was a boy that I met and then I saw him again and he has a den … Mm. I remember there was a dog there and it keep on wanting me to give it some crisps and I keep on having to feed him.
>
> (Anastasia, aged 4, Family I)

Parents'/carers' observations suggest that girls show more general interest in the adult world than boys, picking up on issues being discussed and asking questions in relation to their surroundings, however, this rarely translates into an active interest in drinking:

> Boy: [identifying a picture of a drink] It looks a bit like WKD.
> Interviewer: Right. What's WKD?
> Boy: It's like pop, alcopop, vodka sort of thing, like my sister has every so often …
> Interviewer: Oh right. So it's something that your sister drinks. So when might she have that?
> Girl: When she goes out.
> Boy: When she goes to the pub maybe and at parties
> Girl: [identifying the next picture] It's a cocktail … My Mum and my sister when we go on holiday …
> Interviewer: And do you know any different kinds of cocktails?
> Boy: I've heard of some … Sex on the Beach.
> Girl: I've heard of that one … My Mum and sister had it when we went to Spain.
>
> (Tim and Emma, aged 10, Family C)

When asked about their probable attitudes towards alcohol when they are adults, the children interviewed anticipated a future of moderation. In particular, their imagined futures hinted at a sophisticated recognition that drinking alcohol is a pleasurable and social activity, while also showing awareness of some of the social risks associated with excess consumption, despite their generally limited or confused understanding of the possibilities of alcohol-related harms to physical health.

Conclusion

This chapter has focused on children's knowledge and understanding of alcohol because pre-teen children's knowledge of alcohol has been subject to limited research to-date. Indeed, children have only been addressed indirectly in the second National Alcohol Strategy for England and Wales (2007) as the responsibility of parents/carers/families, rather than as an audience in their own right. Yet, the evidence of this study is that while parents/carers are ambivalent about talking to pre-teen children about drinking, regarding them as too young for such discussions, the children themselves have developed a competent understanding of alcohol and the broad circumstances under which children and adults may drink, as well as having thoughtful reflections about consumption practices of family members and their own likely future behaviour.

Much of this knowledge about alcohol has been gleaned by children through proximal processes, namely their daily interactions with parents/carers/older siblings in the context of everyday family life and from the media, rather than through health campaigns targeted at them or interventions at school. Indeed, children's identification and affective ties with their parents/carers intensifies their learning about alcohol such that their knowledge about different products, where alcohol can be purchased, why people drink, and the social rituals associated with drinking are largely confined to familial consumption practices. In other words, the proximity effect of shared family life produces a particular type of knowledge about alcohol, with the majority of the children in this study, describing positive associations with drinking (e.g. family, friends, shopping, fun). Thus, given parental concerns that pre-teen children are too young to be formally taught about alcohol children's knowledge of the harms associated with drinking are primarily gleaned from television and observation of drunken strangers in public space. The spatiality of children's moral distinction between the meaning of drunkenness at home (silly, makes you sleep), compared with public space (frightening, violent) shows that there is a potentially problematic disassociation between children's understandings of the potential negative effects of drinking to excess and everyday family practices.

Chapter 5
'Do as I Say, or Do as I Do?'

In this chapter we contribute to recent theoretical and empirical work on social reproduction, adult-children interaction, materialities and intergenerational transmission of consumption cultures that are bound up with social rather than individualized notions of consumption. By adding non-representational theories relating to emotions, embodiment and affect to such agendas we also point to new fruitful avenues for research on childhood, families and alcohol, drinking and drunkenness.

'Drinking is for Grown-ups'

Evidence introduced in the previous chapters mirrors Riston's research from the 1970s where most children in our study had tried alcohol either on their own initiative or through parental instigation but either actively disliked the taste of alcohol or preferred soft drinks. Despite children's ambivalence, parent's nonetheless had clear ideas of 'drinking milestones' for their children (see Table 2.1) highlighting how parents/carers adopt dynamic and oppositional stances that define children of different age-genders in relation to each other. Parent's introduce children to adult drinking cultures in order for them to gain a sense of competent conduct in practices of consumption ensuring that 'drinking milestones' were not adhered to in daily life (Jayne et al. 2012). For example as noted in previous chapters, parents/carers/family members encouraged children to engage with adult drinking either through allowing children to 'sip' alcohol or encouraging children to imitate adult drinking practices or rituals albeit with non-alcoholic drinks. This included giving children non-alcoholic cocktails, teaching children how to drink shots of water, and asking children to fetch alcohol at home or to serve alcohol to visitors:

Girl: I think a few times I had one on a Friday or Saturday night when I've actually been sat watching TV with my mum; she's gone to get herself a bottle of wine and she's brought me a little bottle.

(Emma, aged 10, Family C)

As Martens et al. (2004: 174) suggest the ways that 'children engage with objects is instructive for revealing how norms and conventions surrounding material culture are constructed and negotiated between social actors and in terms of how objects script practices during the process of consuming'. Despite children's ambivalence about trying alcohol, parents'/carers' pursuit of 'supervised introduction' to alcohol can be understood as 'anticipatory action' through which children's future lives are 'made and lived in terms of pre-empting, preparing for, and preventing threats to liberal-democratic life' (Anderson 2010: 8–9). Parent's attempts to teach their children about alcohol consumption in the context of home and family lives thus represent 'futures felt as realities which can be used to act upon in the present' (Evans 2010: 21).

However, the overwhelming response by children to the smell of different types of alcohol was either to associate them with other areas of their experiences, or just to consider them as unpleasant:

Boy: Bubblegum.
Interviewer: Bubblegum, like a bubblegum smell?
Boy: Yeah. It doesn't smell of alcohol though, it tastes like one of the drinks that I had at the Scouts last week.

(Tim, aged 10, Family C)

Girl: It reminds me of tooty-fruity flavoured ice-pops.
Interviewer: And you like tooty-fruity ice-pops?
Girl: Yeah because they're all the flavours.
Interviewer: Okay, let's take the lid off. Have a little smell of that one?
Girl: Urgh, I don't like that one, it reminds me of beer [edit] Urgh. It is beer [edit] Yeah, phooey.
Interviewer: So you don't like the smell?
Girl: No.
Interviewer: Tell me what wine is first?

Girl: It's a drink that Mummy and Daddy drinks. [edit] Every time Mummy and Daddy have wine, I hold my nose [edit] its disgusting …

(Lucy, aged 7, Family I)

This evidence points to the significant disjuncture between children's knowledge and interest regarding alcohol, and parents/carers attempts to socialise them into adult drinking cultures. Such conflicts and tensions relating to alcohol consumption and family life highlight how 'social constructions of 'childhood' can be revealed through an analysis of the anxieties surrounding children's consumption' (Martens et al. 2004: 174). In particular we have shown differences between 'model characteristics' of adults attempts to socialise children' into the world of alcohol consumption and children's ambivalence and at times misunderstandings of the pleasures and dangers of drinking. This highlights how 'children's early encounters with the material world are useful for understanding the processes through which that world becomes a meaningful part of culture and, in turn, how that meaning is shaped by the properties of material good in the social world (such as the world of family, friends or school) in which they are put to use' (Martens et al. 2004: 173). Moreover, evidence of children's (mis)understanding of alcohol suggests that passing information/experiences to facilitate *what parent/carers considered* to be a transmission of 'responsible' drinking cultures to their children, could be described as 'self-indulgent attention to the everyday, which never stops to think that children's appreciation of everydayness might be *different*' (Horton and Kraftl 2006: 72).

'Because You'll Walk Funny and You'll be All Confused, then You'll Fall on the Floor'

In the remainder of this chapter we explore how emotions, embodiment and affect offer insights into differences between children and adult appreciation of everyday practices. Specifically, advancing arguments relating to 'proximity' made in Chapter 4 we highlight 'more-than-representational' understanding of childhood and family life with regards to 'liminal thresholds' and 'children's agency in the very moment that children themselves are learning about and coming to grips with the constraints and possibilities of the very different structured environments they encounter in their everyday lives' (James and

Prout 1997: 46). For example, as noted previously most of the children in our study understood what a drunken person looks like and how they behave. The children described 'drunken' adults 'looking tired', with their 'eyes half closed', who 'smell of drink', 'got dizzy' and 'fell on the floor'. These insights show how children were aware of drunkenness as embodied performance as well as 'felt intensity' bound up with adult drinking (Jayne et al. 2010: 55).

Moreover, while children believed that 'drunkenness' is an adult practice, they were aware alcohol consumption would be of interest to 'older children' or teenagers. However, despite children's acknowledgement of adult/ teenagers interest in drinking and drunkenness, our qualitative findings point to clear disassociation between negative social effects of drinking to excess and everyday family practices:

> Boy: Because if kids can get easily drunk and can get drunken quicker [edit] That means you just go all mad and you go like ... you see like lots of ... if you're looking at people, you see lots and people and you go mad in the head. Like you walk over the road thinking you're walking down the street and you could get hit by a car.
> Interviewer: And if I asked you to do an impression of a drunk person .
> Boy: I'll do it ... (does funny walking and makes noises)
>
> (Tim, aged 10, Family C)

> Boy: I think Ant and Dec got drunk or someone like that. And then Ant or Dec thumped the other one, I don't know.
> Girl: They were in a fight.
> Interviewer: Did you read about that somewhere or see it on TV?
> Boy: The Sun, yeah.
> Interviewer: Ah right, they got into a fight.
> Boy: In an elevator.
>
> (Emma and Tim, aged 10, Family C)

> Girl: Well they're a bit like ... sort of like hyper type of thing ... well not like hyper but sometimes they get quite dizzy and they do things that they won't usually do, things like that, yeah.
>
> (Emma, aged 10, Family C)

Children's experiences of drunken adults and older children drinking, highlights an aware of the 'liminal status' [of] … consumption objects and practices' (James and Prout 1996: 61) despite their generally positive responses. For example, children from five of the case study families describe having seen a parent or sibling drunk on occasions commonly related to parties, special occasions or holidays:

> Boy: Yeah, yeah, like weddings, yeah and I think yeah, there has been quite a lot of alcohol and people would get quite drunk. [edit]Yeah, like I would expect like a few of family members to get drunk for weddings and all that, quite big events.
>
> (Kevin, aged 14, Family F)

> Girl: Sometimes Daddy gets drunk.
> Interviewer: Is he funny when he's drunk?
> Girl: No! He always be's sick.
> Interviewer: He's always sick. So who gets drunk the most, Mummy or Daddy?
> Girl: Mummy because she goes out with her friends. [edit] Mostly when they go out like for a cocktail party like my mum did for her birthday and she drunk too much cocktails and she was sick, all over (laughs).
> Interviewer: what are cocktails?
> Girl: Like fruits mixed with wine and alcohol.
> Interviewer: can you have them?
> Girl: No.
> Interviewer: can you have the ones without alcohol in?
> Girl: Yeah, you can have them. [edit]You get like fancy glasses … I've got a cocktail book. [edit] We have them on holiday.
>
> (Lucy, aged 7, Family I)

> Boy: Strange [edit] Have a little bit of a laugh with them, depends how drunk they are. If they're really drunk, stay away. They might go (makes being sick noise).
> Interviewer: And does anyone tell you to stay away or do you just not want to go near them?
> Boy: Just don't want to.
> Girl: At my sister's like 18th, there was this girl who we know, she goes to our school and I think she's 14 and before the party, her mum had took her out

drinking and then she'd gone to the party and she was like really drunk. And then she couldn't walk and she was like really sick.

Boy: Yeah, on the back yard.

(Emma and Tim, aged 10, Family C)

Such evidence points to the ways children correlate 'information': facts, signals, rumours, news, mixed in with moods and emotional energies [allowing them to make sense of an 'adult' activity that is] individual and as an element of a collectivity' (Venn 2010: 132). In these terms our empirical evidence highlights significant problems in parents/carers focus on teaching their children about 'practices' when seeking to transmit 'responsible' drinking cultures across the generations in a manner that ignores children's understanding and experiences of emotions and affective atmospheres that are bound up with adult drinking. While parents/carers were keen to introduce their children to the world of adult drinking by supervising first experiences of tasting/drinking alcohol as well as witnessing 'drunkenness' they were unreflective with regards to the embodied, emotional and affective atmospheres circulating around alcohol consumption with friendship, sociality, closeness, fun, and 'special' family space/times.

For example, returning to an example introduced in previous chapters, holidays are a good example of family space/time that children recognise parents/carers/and/or siblings are most likely to be drunk. In particular, children identified holidays as times when they had observed their parents/carers and/or siblings drunk. Holidays are celebrated as times when people can relax by forgetting their normal routines/rules and by indulging themselves in ways that they would not normally do so at home (e.g. by staying up late, dressing differently, and eating and drinking more than usual):

Mother: I mean when we've been away on holidays, the girls have had you know, Bacardi Breezer which is very low in alcohol ... I suppose I feel that sometimes if you're not allowed to do it, that's where you push boundaries and you'll do it.

(Parents, Family G)

Boy: Are those shots?

Girl: I've seen shots ...

Interviewer: So do you know what's special about shots or what's different?

Boy: Is it strong vodka or something like that? ...

Girl: Yeah because it's only small … so you could probably drink three and be drunk.
Interviewer: And have you ever seen anyone actually drinking them?
Girl/boy: Yeah.
Interviewer: And how do they do it?
Boy: They go like that, in one …
Girl: I've seen my Mum do it when we went to Greece because we did this quiz one night at the bar and our team won and they got free shots.

(Tim and Emma, aged 10, Family C)

Indeed, the comments highlight that children were well aware that the 'affective space' of holiday atmospheres, and embodied and emotional relaxation and fun was often achieved through alcohol consumption (Valentine et al. 2012):

Boy: Well I've never seen someone like … I've seen someone really happily drunk, like sitting round at a table and laughing a lot but not like on the floor like being sick or all that, I haven't seen that.

(Kevin, aged 14, Family F)

Children were thus clear that alcohol consumption is 'not merely the private, subjective enjoyment of the body, but also a symbolic transformation of feeling with the body … That is to say, they [children were aware that drinkers] inhabit an imaginary world of their own making, central to which is their comportment and that of their fellows' (Radley 1995: 11). Yet, none of the parent's recognised holidays as a time when they were modelling specific, albeit different from their 'normal', drinking practices. In this sense, because 'the holiday' is imagined as a space outside of everyday practices they were not perceived to have a significant impact on children's attitudes towards, or future practices of, alcohol consumption and is therefore just like 'the home' – a family space/time where children can be exposed to adult alcohol consumption in a 'safe' manner.

Our findings relating to 'drunken' adults on holiday or at parties, or older children and teenagers also directly relate to James and Prout (1996: 49) notion of liminality representing a 'furnace of consumption practices aimed at mediating this disruptive, unstable period of time'. Moreover, as noted in previous chapters children in our study were not threatened or upset by their parents'/carers' drunken moments. Rather, they commonly represented their parents'/carers' uncontrolled or silly behaviour in a bemused way, although one child recognised degrees of drunkenness and that if someone is 'a bit drunk' you can have fun with them but if 'very drunk' you should stay away. Such evidence shows that children in our study recognized 'that alcohol, drinking and drunkenness is used [by adults] to generate particular embodied, emotional and affective experiences that, while varying between individuals, nonetheless creates collective drunken mind-body-worlds … enabling interaction with people, space and place' (Jayne et al. 2010: 551).

Discourses relating to dangers of alcohol consumption, which tended to contrast with our adult respondents own experiences of the pleasures of drinking in public spaces and commercial venues are akin to Cody's (2011: 61) depiction of 'fruitful darkness' which 'encapsulates the concurrent darkness and energy, the restorative obscurity that epitomizes … experiences within the intricacies of social categorization'. Indeed by recognizing the emotional, embodiment and affective space/times of adult drinking children show a 'sensitivity to practices might help us to see the world somewhat anew; to realize how many ostensibly inevitable, common sense or fixed things in the world are actually always-on going and constantly reformulated in and through practice' (Horton and Kraftl 2006: 75). While many of the older children in our study did nonetheless recognise that adult's drink alcohol to have fun and to relax as part of their everyday lives; in a manner that could be positive for both adults and children, the younger children generally viewed parental and adult drunkenness as 'out of the ordinary' or were ambivalent to adult drinking practices altogether. Children are thus aware that alcohol consumption allows responsiveness to and openness towards the worlds of others, involving an interweaving of the personal with the social, and the affective with the mediated offering opportunities for drinkers to engage with friends and strangers in a way that they may not do if they haven't been drinking alcohol.

Despite such insights however adult/carers paid little attention to the ways in which children recognize both individual and collective emotions, embodiment and affect that are bound up with alcohol, drinking and drunkenness. Anderson's (2009: 80) contention that 'atmospheres are a kind of indeterminate affective 'excess' through which intensive space-times can be created' highlights the problematic way in which the home/holiday as a space/time where parents/carers are happy to introduce children to the pleasures and dangers of alcohol in a 'safe' environment are discursively constructed and represented by parents/carers in opposition to public spaces where they associate alcohol consumption with violence and disorder (Valentine et al. 2010). This emphasis over-simplifies public and domestic drinking cultures and misses the opportunity to teach children about the range of other drinking practices and spaces that they may encounter throughout their lives (Jayne et al. 2012).

Conclusion

In this chapter we have illuminated relationships between childhood, family life and alcohol with reference to emotions, embodiment and affect. Focusing on the ways in which parents'/carers' seek to reconcile the 'liminal' status of alcohol consumption through attempts to ensure inter-generational transmission of pleasures and dangers of alcohol consumption have been shown to conflict with pre-emptive approaches to teaching children about dangers relating to drinking. Our research foregrounds relationships, obligations and reciprocity bound up with consumption practices and identity formation to offer insights into the everyday life of children and adults. In doing so we highlight social reproduction, adult-children interaction, and intergenerational relations that are discursively and differentially constructed as contradictory and expressed through lived experiences and materialities relating to alcohol consumption as both a signifier of both social harm and as indicative of productive family and social relations.

More specifically, we have shown that engaging with embodiment, emotions and affect offers new ways in which theoretical and empirical understanding of relationships between childhood, family life and alcohol, drinking and drunkenness can be advanced. Theorising and researching childhood, families and alcohol as signified, articulated and 'felt' highlights individual/collective 'subjects which are caught and situated as bodies with radiating ripples and circuits of feeling, intensity, response and sensation … [via] flows that wrap into and fold out of our bodies' (Dewsbury 2009: 21). Fun, excitement, enchantment, celebration, relaxation, tension and conflict thus offer fruitful avenues for understanding how consumer cultures are reproduced and transformed through the life cycle and across generations. Sustained engagement with emotions, embodiment and affect clearly have significant contributions to make in advancing understanding of social/individualized consumption with reference to social reproduction, adult-children interaction, materialities, and intergenerational 'transmission' of drinking cultures.

Chapter 6
Conclusions

Throughout this book we have shown that the dominant attitude of parents/ carers surveyed and interviewed for our research was that children should be introduced to alcohol by families at home. Parents/carers who drank wanted their children to appreciate the benefits and pleasures of alcohol as well as the risks associated with drinking to excess and aspired to teach their own children to drink in moderation in adulthood. Indeed, the case study research with children found that they had absorbed this message, recognised that alcohol is an adult product (including age restrictions on the sale of alcohol in supermarkets), were aware of the social albeit not the health harms of drinking to excess and anticipated that in their future lives they would only drink in moderation. However, it is impossible to know whether (and if so, how) the children's understanding of alcohol will translate into their adult lives without undertaking qualitative longitudinal research.

Most of the parents/carers who participated in the case study element of the research did not have specific rules and guidance related to alcohol for children aged 5 to 12 as they did not consider them to be interested in drinking at this age. At the same time, the case study parents/carers modelled a positive attitude towards alcohol – emphasising pleasure and sociality (notably a reverse morality that good behaviour can be rewarded by a 'naughty but nice' drink) – through their domestic drinking (and shopping) practices, including encouraging children to try alcohol and to participate in drinking rituals, albeit often by mimicking these practices and moralities with soft drinks at a much earlier age than current government guidelines recommend. However, the parents/carers were more reluctant to expose children to drinking (unless it was with a meal) in public spaces, thus implicitly constructing an understanding of home as a safe space to drink in contrast to the risks associated with alcohol in public spaces, a message children further absorbed from media representations of 'public' drinking in both news and drama programmes.

Indeed, those who took part in the qualitative element of the research had an individualised approach to parenting about alcohol. They were reluctant to reprimand other people's children for inappropriate drinking or for other adults to discipline their own children; resistant to the suggestion that alcohol education should be provided at school; and considered that general advice in relation to alcohol would be ineffective, believing that each child had an individual personality and needed to be parented in specific ways. In emphasising children's expressivity, rather than parental discipline, the interviewees presented families as resources out of which individual children construct themselves, defining the role of parents/carers as to equip children with the right personal qualities and skills to ensure that they make sensible choices in relation to alcohol. This neoliberal model of parenting assumes that a child is able to distinguish between what might be the right action for himself/herself in a particular time and place. However, it does not recognise that an individual's drinking can impact on many other lives beyond his/her own and consequently it does not acknowledge the wider shared social responsibilities of adults or children. Yet, as some parents/carers observed when complaining about other parents/carers allowing their children to binge drink in public spaces – without appreciating the significance of their comments – not all children have positive family support and as a consequence some children are much less well equipped to make 'sensible choices' than others.

The implications of these findings suggest that government advice such the former Chief Medical Officer's (2009) recommendation that children under 15 should avoid alcohol completely appears to be unrealistic given that alcohol is an unremarkable and taken for granted part of many families' everyday lives. There is a danger that such advice 'problematises' what appear to be sensible parental attitudes and approaches to alcohol. Indeed, at ages 5 to 12 it is parents/carers who are the most important influences on children's attitudes towards alcohol. Contemporary parenting strategies appear to be largely successful at conveying the social pleasures and risks of drinking alcohol at home, and the message that alcohol should be consumed in moderation. Indeed, young children even appear to learn positive messages about moderation from witnessing their parents/carers/relatives drinking to excess. However, the findings of this research suggest that there are gaps in younger children's knowledge and understanding around alcohol use, which have implications for those bodies that currently provide guidance to parents/carers about how to talk to children about alcohol (e.g. government departments, service providers, voluntary, charitable and independent organisations).

The children in this study did not appear to be being taught to recognise the potential health implications of alcohol consumption. The research suggests that some parents/carers are unaware of the current sources of advice or are reluctant to/ineffective at transmitting health messages to their children. The social risks associated with alcohol readily arise in many households because of intra-familial practices (e.g. parents'/carers' modelling behaviour), as well as the visibility of popular and political debates about alcohol (e.g. on television). In contrast, the public health risks (defined by the Department for Education as cancer of the mouth and throat, sexual and mental health problems, liver cirrhosis and heart disease) do not resonate with parents'/carers' own experiences of alcohol and are less easily raised in the context of everyday family life. In addition, by focusing on the social risks associated with alcohol the parental emphasis on learning to drink at home in a 'safe' environment over-simplifies and misses the opportunity to teach children about the range of other drinking practices and spaces they may encounter throughout their lives. It appears that the parental stress on individual choice also does not educate children about the impact their personal drinking and occasional drunken behaviour may have on others and consequently their potential wider social responsibilities as consuming adult citizens of the future.

In addition to these gaps in what parents/carers are teaching children it is also important to recognise that not all young people have the familial support described throughout this book. For example, some may be over-exposed to 'problem' drinking, others 'over-protected' from knowledge about alcohol, or not informed about alcohol for cultural or religious reasons given that there is some evidence of the presence of alcohol even in communities that abstain (Braby 2007; Valentine et al. 2010b). Alcohol education is therefore one way to address the gaps in what children are learning about alcohol and the differential levels of education and support children receive at home. Indeed, the Department for Education states that children aged 7 to 11 will learn about the health and social risks associated with alcohol and the basic skills for making good choices about their health and recognising risky situations at school as part of the National Curriculum. Yet the majority of children who participated in this study stated that they had not been taught about alcohol at school, which suggests that this education is either not taking place, or is not being delivered in an effective manner. The findings of this study imply that it would be beneficial for the Department for Education to review the way alcohol education is currently delivered as part of the National Curriculum (for 7- to 11-year-olds) within primary schools in order to improve its efficacy. As part of this, schools should be encouraged to involve parents/carers in order that the same key messages about alcohol can be reinforced at both home and school. To maximise impact, any alcohol education in schools should be run in parallel with campaigns targeted at parents/carers.

Childhood, Family, Alcohol: Making a Difference?

The empirical research and findings presented throughout this book has been underpinned by four overlapping theoretical debates that we have drawn together in order to better understand childhood, family life and alcohol, drinking and drunkenness. Firstly, we highlighted the importance of undertaking a cross-generational perspective by exploring pre-teen children's understandings of alcohol as well as that of parents/carers. Secondly, we focused on the significance of the spaces of everyday family life: understanding childhood in context. Thirdly, we argued for the importance of considering adult alcohol consumption, drinking practices and drunken performativities in relation to children and childhood in order to offer useful insights into recent debates abound social/individualized consumption, focused on social reproduction, adult-children interaction, materialities, and intergenerational 'transmission' of drinking cultures. And finally, we showed the importance of non- (and more-than) representational understanding of complexities of childhood, family life and alcohol consumption.

These theoretical foundations were vital in ensuring that our findings were relevant to partnership work with policy makers and practitioners. Our funding was provided by the Joseph Rowntree Foundation (JRF), and enabled collaboration with a diverse range of partners including Alcohol Concern and the Drinkaware Trust (see Valentine et al. 2008; 2010). Importantly the research took place with a backdrop of new imperatives and opportunities to assess the *impact* of policy relevant theoretically and empirically robust academic work. In the UK, the 'impact agenda' emerged from previous attempts to map what had been known in various guises as 'academic enterprise' or 'policy and knowledge transfer'. For example, under the periodic review of academic research in the UK, the Research Excellence Framework 2014 (previous years known as Research Assessment Exercise) now requires collation of *evidence* of *impact* defined as:

> an effect on, change or benefit to the economy, society, culture, public policy or services, health, the environment or quality of life, beyond academia … [edit] Impact **includes**, but is not limited to, an effect on, change or benefit to: the activity, attitude, awareness, behaviour, capacity, opportunity, performance, policy, practice, process or understanding of an audience, beneficiary, community, constituency, organisation or individuals; vin any geographic location whether locally, regionally, nationally or internationally … [edit] Impact **includes** the reduction or prevention of harm, risk, cost or other negative effects.
>
> (HEFCE 2012: 48)

In seeking to measures the impact of our ongoing research into geographies of alcohol, drinking and drunkenness we found that our research findings had been utilised by: research organisations (Joseph Rowntree Foundation, Royal Geographical Society), campaigning groups (Alcohol Concern, Quaker Action on Alcohol, Drugs, Family and Alcohol Alliance) and trade-led bodies (Drinkaware Trust, Portman Group). Our research had also fed into national policies and on-going programmes, including the UK Government Youth Alcohol Action Plan (2008), and the House of Commons Health Committee inquiry (and report) on the Government's Alcohol Strategy (2012). Our keynote lectures at policy conferences and workshops attracted key actors, organisations and institutions involved in policy and practice.

We were also able to gather testimonies from our partners that our research had impacted on the following policies, practices, initiatives and campaigns: inspiring the Joseph Rowntree Foundations *Alcohol Programme* to foregrounding geographical approaches to understanding alcohol consumption. Moreover, drawing on our research JRF worked closely with the alcohol strategy/policy teams in Westminster, Scotland, Northern Ireland. We also gathered statistical based evidence to show how our research on childhood and family life (Valentine et al. 2010a) had directly influenced the content of the Drinkaware Trusts campaign 'Your Kids and Alcohol'; especially the production of an online video which was viewed over 800,000 times, as well as the content of an advice leaflet which was requested by, and delivered to 100,000 families. The Drinkaware trust also significantly drew on the research into alcohol, families and childhood to develop In:tuition, a course of 10 primary lessons and 11 secondary lessons aimed at building young people's confidence, decision-making and communication skills. To date, 751 schools and 909 other organisations (PCTs, Youth Services, local authorities, etc.) utilized this resource. In:tuition was then rolled out 650 schools in the UK from 2013–14. We also collected evidence showing that our research also generated strong traditional and social media interest and debate.

We are sure that there are many other individuals and research groups who apply social and cultural theory to studying alcohol, drinking and drunkenness that have similarly uncovered impressive evidence of the impact of their research. However, such comments notwithstanding, ethnographic (and in this case mixed-methods) research has failed to significantly impact on the dominance of scientific research; lab-based experimentation, statistical measurement, modelling and numerical proxies which continue to define alcohol-related harm and related policy and practice. Moreover, 'alcohol studies' across the social and medical sciences is defined by an 'impasse where alcohol consumption is conceived as a medical issue, pathologised as a health, social, legislative, crime or policy problem *or* as being embedded in social and cultural relations – with limited dialogue between these approaches' (Jayne et al. 2008a: 247). With this backdrop it remains scientific and quantitative findings that generally capture newspaper headlines, dominate popular imaginations and representations and influence political decision-making and policy formation (see Jayne and Valentine 2015, 2016). While a lack of progress in challenging ontological and epistemological orthodoxies is disappointing, as we have shown in this book, theoretically informed ethnographic research nonetheless offers significant opportunities to generate a rigorous evidence base that can impact on academic, popular, political and policy debate. In doing so, we have outlined new theoretical and empirical terrain which offers fruitful possibilities to advance understanding of the complexities, contradictions and nuances bound up with childhood, families and alcohol, drinking and drunkenness.

Appendix 1:
Research Design

The research in this book is based on a multi-method research design. Specifically, we employed a representative survey to establish national patterns in relation to parents'/carers' attitudes and practices towards the role of alcohol within the family, and in-depth case studies, involving interviews, participant observation and child-centred methods, to unpack the specific processes through which drinking cultures are transmitted/interrupted within families.

National Survey

The survey was designed and piloted by the research team and conducted by a social research company. It was implemented by telephone to a nationally representative sample of 2,089 parents/carers with at least one child aged between 5 and 12. Appendix 2 summarises key demographic and socio-economic characteristics of the respondents. The representativeness of the social economic status of respondents was checked against both the 2001 Census and 2005 Labour Force Survey (ONS 2004; Hall 2006), and the representativeness of the educational status of the respondents was verified by comparing information on the highest level of qualification: by marital status and presence of dependent children drawn from the 2008 Labour Force Survey (ONS 2009).

The survey was organised with five sections:

- household and family structure;
- family life and parenting (general);
- family life, parenting and alcohol;

- family life and alcohol (actual);
- demographics.

Specifically, we collected data on the parents'/carers' perceived (above/below/ at recommended levels) and actual alcohol consumption practices (what they drink, where, when, with whom and how this relates to the presence of children); their perceptions of national/local norms in relation to attitudes towards the role of alcohol in the family (how this varies by children's age/ gender and in relation to their own childhoods); and their awareness of the law and perceptions of national/local 'norms' in relation to children and alcohol. The sampling strategy allowed us to establish national patterns in relation to parents'/carers' attitudes and practices towards the role of alcohol within the family, and to benchmark the qualitative case study research within this national context. It was also used to recruit family participants for the qualitative research phases.

Case Study Research

Ten families, with at least one child aged between 5 and 12, were recruited via the survey as case studies for the multi-stage qualitative research (cf. Wallman's 1984 *Eight London households*). The case studies were purposively sampled on the basis of the survey results to include families with diverse structures, socio-economic profiles and a range of attitudes and practices to drinking (including those who drink dangerously above, within and below safe limits; see Appendix 3). We did not interview households that abstain from alcohol. Each family took part in five activities.

Family interviews/social network analysis: Interviews with parent(s)/carer(s) collected data about their attitudes/practices towards parenting, specifically in relation to alcohol in the family. These included discussion of the parent(s)'/ carer(s)' awareness of how their own behaviour, and that of other family members, might influence their children's current/potential/future drinking practices and their deliberate strategies (and associated communication styles) for transmitting sensible drinking practices to their children. We also explored parent(s)'/carer(s)' experiences of dealing with children's misuse of alcohol or possible strategies to deal with future problems; differences in their attitudes/practices towards children according to age/gender/position in the family; and differences between their own attitudes/practices/communication styles as mothers/fathers and the complexities of how these relationships are negotiated/resisted/contested within the family (by children/other significant adults).

Where families were constituted by two parents/carers they were interviewed together where possible. As part of the interview process the parent(s)/carer(s) completed a social network analysis form. This identified the adults (relatives, friends, neighbours, other parents/carers etc.) who play an important part in the family's life and influence the mother's/father's practices in relation to children's exposure to alcohol, and provides a measure of the social capital available to support/develop sensible drinking habits.

The children's experiences of the above issues were explored through a child-centred interview process that as well as 'conventional' interview-style questions included exploring children's understanding of alcohol by asking them to identify samples of drinks (alcoholic and non-alcoholic) by smell, and asking children which drinks they could identify from a series of advertisements for common products/brands. The role of alcohol in the family was then explored with the youngest children by using puppets or a dolls house to play-act a family party. Older children were shown clips from episodes of the cartoon series *'The Simpsons'* which represent both adults and children as drinking/drunk. These were used as a basis for a wider discussion about their attitudes to alcohol and family practices. Where siblings were of similar ages they were 'interviewed' together, but where there was a significant age gap the interviews were conducted separately.

Participant observation at a celebration: Families invited a member of the research team to a special event where alcohol was consumed (e.g. birthday/anniversary/fireworks party, wedding etc.). This participant observation involved *descriptive observations* (about the location, guests, general activities/specific acts, ambience, role of children and their interaction with adults in relation to alcohol etc.) and *narrative accounts* that built up an overall picture of how attitudes/practices relating to alcohol were transmitted to children within the context of such events.

Individual interviews with parent(s)/carer(s): These interviews explored the individual parent(s)'/carer(s)' own childhood experiences of alcohol, how they were parented and how these experiences have influenced their individual attitudes/practices towards parenting, specifically in relation to alcohol in the family. These individual interviews offered the opportunity to investigate perceptions of any conflicts/tensions/negotiations around alcohol that occur between mothers/fathers/otherparent(s)/carer(s) as well as between the individual parent/carer and their children/siblings.

Photo-elicitation activity and participation observation of a family meal: This element allowed consideration of drinking at a 'special' family meal compared with a 'normal' family meal. Participants were asked either to photograph an eating/drinking event related to an appropriate religious or cultural festival (e.g. Christmas) or to dig out family photographs of previous such events. These images were used in a family interview to discuss how drinking in 'special' time/space contexts promotes or interrupts the transmission of everyday family drinking practices to children.

Participant observation of a 'normal' family treat involving alcohol: The researcher accompanied the family on a 'normal' treat that involved alcohol (e.g. meal out, a sporting/leisure/entertainment event, shopping etc.). The participant observation involved both descriptive observations and narrative account about children's interaction with adults in relation to alcohol. This also acted as a 'farewell' event where the family reflected on the research process and discussed the emerging findings with the researcher.

Appendices

Appendix 2: Characteristics of the Survey Respondents

Variable	Data summaries
Gender	Female = 1,535 (73.5%); male = 554 (26.5%); missing = 0
Age	Mean = 39.6, median = 40; minimum = 21, maximum = 69; inter-quartile range = 35–44; missing = 4
Marital status	Married = 1,577 (75.5%); Cohabiting = 147 (7%); single never married = 138 (6.6%); divorced = 106 (5.1%); separated = 85 (4.1%); widowed = 19 (1%); other relationship = 17 (0.8%)
Ethnic group	White UK = 1,857 (89%); white other = 94 (4.5%); black = 64 (3.1%); Asian = 54 (2.6%); mixed/other = 20 (1%)
Religion	Christian = 1,284 (61.6%); none = 674 (32%); Muslim = 46 (2.2%); other religion = 40 (1.9%); Hindu = 19 (0.9%); Jewish = 15 (0.7%); refused = 11 (0.5%)
NS-SEC	Lower managerial/professional and intermediate = 784 (37.5%); own, small and lower supervisory/technical = 487 (23.4%); semi-routine and routine = 325 (15.5%); large employers and higher managerial/professional = 286 (13.6%); unemployed and refused = 207 (10.0%)
Highest education level	Below A-level = 540 (25.8%); NVQ 4 and 5 and other vocational = 430 (20.6%); first degree = 392 (18.8%); A-level = 358 (17.2%); higher degree = 222 (10.6%); no qualifications = 147 (7%)
Don't drink	415/2,089 = 20%
Interested in participating in further research	859/1,674 = 51.3%

Appendix 3: Definition of Binge Drinking and Guide to Alcohol Unit Measurements

In 1995 the UK Government report *Sensible Drinking* changed the guidelines for recommended limits from a weekly to a daily measure of consumption, reflecting the concern that: '*weekly consumption can have little relation to single drinking episodes and may indeed mask short term episodes which … often correlate strongly with both medial and social harm*'. The change from an emphasis on weekly to daily levels does not increase the recommended upper limit for weekly consumption.

The current Department of Health advice is that men should not drink more than 3–4 units of alcohol per day, and women should not drink more than 2–3 units of alcohol per day. Binge drinking is less clearly defined, but has been referred to be the Department of Health and Office for National Statistics as '*consuming eight or more units for men and six or more units for women on at least one day during the week*'. In other words double the daily recommended levels of consumption.

One unit of alcohol is measured as 10ml of pure alcohol and guidance is given in terms of particular drinks. As a rough guide, the following unit measurements apply:

- A pint of ordinary strength lager: 2 units
- A pint of strong lager: 3 units
- A pint of bitter: 2 units
- A pint of ordinary strength cider: 2 units
- A small (175ml) glass of wine: 2 units approx
- A measure of spirit: 1 unit

- An alcopops: 1.5 units approx

It is however very difficult to be accurate as measures, strengths and types of alcohol vary considerably.

Bibliography

Ahmed, S. (2004) 'Collective feelings or the impression left by others', *Theory Culture and Society*, 212: 25–42.

Alcohol Concern (2007) *Glass Half Empty*, London: Alcohol Concern.

Alderson, P. (1993) *Children's Consent to Surgery*, Buckingham: Open University Press.

Anderson, B. (2009) 'Affective atmospheres', *Emotion, Space and Society*, 2: 77–81.

Anderson, B. and Harrison, P. (2010) *Taking-Place: Non-Representational Theories and Geography*, Aldershot: Ashgate.

Anderson, D. (2010) 'Pre-emption, precaution, preparedness: anticipatory action and future geographies', *Progress in Human Geography*, 34: 777–798.

Anderson, P. and Baumberg, B. (2006) *Alcohol in Europe: A Public Health Perspective*, London: Institute of Alcohol Studies.

Andersson, B. and Hibell, B. (2007) 'Drunken behaviour, expectancies and consequences among European students', in Järvinen, M. and Room, R. (eds) *Youth Drinking Cultures: European Experiences*, Aldershot: Ashgate, pp. 41–64.

Andrews, J., Tildesley, E., Hops, H., Duncan, S. and Severson, H. (2003) 'Elementary school age children's future intentions and use of substances', *Journal of Clinical Child and Adolescent Psychology*, 32: 556–567.

BBC (2015) 'Binge drinking continues to fall in young adults' http://www.bbc.co.uk/news/health-31452735 (accessed 16/02/2015)

Beck, U. and Beck-Gernsheim, E. (2002) *Individualisation*, Cambridge: Sage Publications.

Beck, U. and Beck-Gernsheim, E. (1995) *The Normal Chaos of Love*, Polity Press: Cambridge.

Berch, D., Haguquist, C. and Starrin, B. (2011) 'Parental monitoring, peer activities and alcohol use: a study based on data on Swedish adolescents', *Drugs: Education, Prevention and Policy*, 18(2): 100–107.

Bogenschneider, K., Wu, M-Y., Raffaelli, M. and Tsay, J. C. (1998) '"Other teens drink, but not my own kind": Does parental awareness of adolescent alcohol use protect adolescents from risky consequences?', *Journal of Marriage and Family*, 356–373.

British Medical Association Board of Science (2008) *Alcohol Misuse: Tackling the UK Epidemic*, London: British Medical Association.

Cameron, C., Stritzke, W. and Durkin, K. (2003) 'Alcohol expectancies in late childhood: an ambivalence perspective on transitions towards alcohol use', *Journal of Child Psychology and Psychiatry*, 44(5): 687–698.

Casswell, S., Brasch, P., Gilmore, L. and Liva, P. (1985) 'Children's attitudes to alcohol and awareness of alcohol-related problems', *British Journal of Addiction*, 80: 191–194.

Casswell, S., Gilmore, L. L., Silva, P. and Brasch, P. (1988) 'What children know about alcohol and how they know it', *British Journal of Addiction*, 83: 223–227.

Cody, K. (2012) 'No longer, but not yet: tweens and the mediating of threshold shelves through liminal consumption', *Journal of Consumer Culture*, 12: 41–64.

Coleman, L. and Cater, S. (2007) 'Changing the culture of young people's binge drinking: from motivations to practical solutions', *Drugs: Education Prevention and Policy*, 14(4): 305–317.

Conway, K. P., Swenden, J. D. and Merikangas, K. R. (2002) 'Alcohol expectancies, alcohol consumption and problem drinking: the moderate role of family history', *Addictive Behaviours*, 28: 832–836.

Cook, D. T. (2003) 'Spatial biographies of children's consumption: market places and spaces of childhood in the 1930s and beyond', *Journal of Consumer Culture*, 3: 147–169.

Cook, D. T. (2008) 'The missing child in consumption theory', *Journal of Consumer Culture*, 8: 219–243.

Cox, R. and Narula, R. (2003) 'Playing happy families: rules and relationships in au pair employing households in London, England', *Gender Place and Culture*, 10: 333–44.

Davidson, J., Smith, M. and Bondi, L. (eds) (2005) *Emotional Geographies*, Aldershot: Ashgate.

Department of Health and Home Office (2007) *Safe, Sensible and Social: The Next Steps in the National Alcohol Strategy*, London: Department of Health and Home Office.

DeVerteuil, G. and Wilton, R. (2008) 'The geographies of intoxicants: from production and consumption to regulation, treatment and prevention', *Geography Compass*, 3/1: 479–494.

Donaldson, L. (2009) *Guidance on the Consumption of Alcohol by Children and Young People*, London: Department of Health.

Dorn, M. (1999) 'The moral topography of intemperance', in Butler, R. and Parr, H. (eds) *Mind and Body Spaces: Geographies of Illness, Impairment and Disability*: London: Routledge. pp. 46–69.

Dyck, I. (1996) 'Mother or worker? Women's support networks, local knowledge and informal childcare strategies', in England, K. (ed.) *Who will Mind the Baby? Geographies of Child-care and Working Mothers*, London: Routledge: London. pp. 123–140.

Eadie, D., MacAskill, S. and Brooks, O. (2010) *Pre-teens Learning about Alcohol: Drinking and Family Contexts*, York: Joseph Rowntree Foundation.

Ettore, E. (1997) *Women and Alcohol: A Private Pleasure or Public Problem?* London: The Women's Press.

Evans, B. (2010) 'Anticipating fatness: childhood, affect and the pre-emptive 'war on obesity', *Transactions of the Institute of British Geographers*, 35: 21–38.

Evans, B. (2008) 'Geographies of youth/young people', *Geography Compass*, 2(5): 1659–1680.

Evans, B., Colls, R. and Horschelmann, K. (2011) 'Change4Life for your kids?: embodied collectives and public health pedagogy', *Sport, Education and Society*, 16(3): 323–341.

Fevre, R. W. (2000) *The Demoralisation of Western Culture*, London: Continuum.

Forsyth, A. and Barnard, M. (2000) 'Preferred drinking locations of Scottish adolescents', *Health and Place*, 6: 105–115.

Foxcroft, D. R. and Lowe, G. (1997) 'Adolescent's alcohol use and mis-use: the socialising influence of perceived family life', *Drugs: Education, Prevention and Policy*, 4: 215–229.

Foxcroft, D. R. and Lowe, G. (1991) 'Adolescent drinking behaviour and family socialisation factors: a meta-analysis', *Journal of Adolescence*, 14: 255–273.

Gabb, J. (2010) *Researching Intimacy in Families*, Basingstoke: Palgrave Macmillan.

Gagen, E. (2000) 'Playing the part: performing gender in America's playground', in Holloway, S. L. and Valentine, G. (eds) *Children's Geographies*, London: Routledge. pp. 213–229.

Gaines, L., Brooks, P., Maisto, S., Dietrich, M. and Shagena, M. (1988) 'The development of children's knowledge of alcohol and the role of drinking', *Journal of Applied Development Psychology*, 9: 441–457.

Gillies, V. (2009) *From Function to Competence: Engaging with New Politics of Family*, Paper presented at the Rethinking Concepts, BSA Study Group Colloqium, London.

Gillis, J. (1996) *A World of their Own Making: Myth, Ritual and the Quest for Family Values*, New York: Basic Books.

Greene, S. and Hogan, D. (eds) (2005) *Researching Children's Experience: Approaches and Methods*, London: Sage.

Gullestad, M. and Segalen, M. (eds) (1997) *Family and Kinship in Europe*, London: Pinter.

Higher Education Funding Council for England (2012) *Research Excellence Framework: Assessment Framework and Guide on Submission*, HEFCE: Bristol.

Herrick, C. (2011) 'Why we need to think beyond "the industry" in alcohol research and policy studies', *Drugs: Education, Prevention and Policy*, 18(1): 10–15.

Hibell, B., Guttormsson, U., Ahlstrom, S., Balakireva, O., Bjarnason, T., Kokkevi, A., Kraus, L. (2009) *The 2007 ESPAD Report: Substance Use Among Students in 35 European Countries*, Stockholm: Swedish Council for Information on Alcohol and Other Drugs and the Pompidou Group at the Council of Europe.

Highet, G. (2005) 'Alcohol and cannabis: young people talking about how parents respond to their use of these two drugs', *Drugs, Education, Prevention and Policy*, 12: 113–124.

Holloway, S. L. (1999) 'Mother and worker?: the negotiation of motherhood and paid employment in two urban neighbourhoods', *Urban Geography*, 20: 438–460.

Holloway, S. L. (1998) 'Local childcare cultures: moral geographies of mothering and the social organisation of pre-school education', *Gender, Place and Culture*, 5: 29–53.

Holloway, S. L., Hubbard, P. J., Jöns, H. and Pimlott-Wilson, H. (2010) 'Geographies of education and the importance of children, youth and families', *Progress in Human Geography*, 34(5): 583–600.

Holloway, S. L., Jayne, M. and Valentine, G. (2009) 'Masculinities, femininities and the geographies of public and private drinking landscapes', *Geoforum*, 40: 821–31.

Holloway, S. L., Jayne, M. and Valentine, G. (2008) '"Sainsbury's is my Local": English alcohol policy, domestic drinking practices and the meaning of home', *Transactions of the Institute of British Geographers*, 33: 532–547.

Holloway, S. and Valentine, G. (2000) *Children's Geographies: Playing, Learning, Living*, London: Routledge.

Horton, P. and Kraftl, P. (2006) 'What else? Some more ways of thinking and doing "children's geographies"', *Children's Geographies*, 4(1): 69–95.

Hubbard, P. (2005) 'The geographies of "going out": emotions and embodiment in the evening economy', in Davidson, J., Bondi, L. and Smith, M. (eds) *Emotional Geographies*, Aldershot: Ashgate. pp. 117–134.

Jackson, C., Henricksen, L. and Dickenson, D. (1997) 'Alcohol-specific socialization, parent behaviours and alcohol use by children', *Journal of Studies on Alcohol*, 362–367.

Jahoda, G. and Cramond, J. (1972) *Children and Alcohol*, London: HMSO.

James, A. (1990) 'The good, the bad and the delicious: the role of confectionary in British society', *Sociological Review*, 38: 666–88.

James, A. and Prout, A. (1997) *Constructing and Reconstructing Childhood: Contemporary Issues in the Sociological Study of Childhood*, London: Routledge.

Järvinen, M. and Room, R. (eds) (2007) *Youth Drinking Cultures: European Experiences*, Aldershot: Ashgate.

Jayne, M. and Valentine, G. (2016) 'Drinking dilemmas: making a difference? in Thurnell-Read, T. (ed.) *Drinking Dilemmas: Sociological Approaches to Alcohol Studies*, Routledge: London. (In press)

Jayne, M. and Valentine, G. (2015) 'Alcohol-related violence and disorder: new critical perspectives', *Progress in Human Geography*, (OnlineFirst).

Jayne, M., Valentine, G. and Gould, M. (2012) 'Family life and alcohol consumption: the transmission of "public" and "private" drinking cultures', *Drugs: Education, Prevention and Policy*, 19: 192–200.

Jayne, M., Valentine, G. and Holloway, S. L. (2011a) 'What use are units? Critical geographies of alcohol policy', *Antipode*, 44(3): 828–846.

Jayne, M., Valentine, G. and Holloway, S. L. (2011b) *Alcohol, Drinking and Drunkenness: (Dis)orderly Spaces*, Aldershot: Ashgate.

Jayne, M., Valentine, G. and Holloway, S. L. (2010) 'Emotional, embodied and affective geographies of alcohol, drinking and drunkenness', *Transactions of the Institute of British Geographers*, 35(4): 540–554.

Jayne, M., Valentine, G. and Holloway, S. L. (2008a) 'Geographies of alcohol, drinking and drunkenness a review of progress', *Progress in Human Geography*, 32(2): 247–263.

Jayne, M., Valentine, G. and Holloway, S. L. (2008b), 'Fluid boundaries – 'British' binge drinking and 'European' civility: alcohol and the production and consumption of public space, *Space & Polity*, 12(1): 81–100.

Jayne, M., Holloway, S. L. and Valentine, G. (2006) 'Drunk and disorderly: alcohol, urban life and public space', *Progress in Human Geography*, 30: 451–468.

Jernigan, D. (2005) 'The USA: alcohol and youth today', *Addiction*, 100: 271–73.

Jones, J. (2002) 'The cultural symbolisation of disordered and deviant behaviour: young people's experiences in a Welsh rural market town', *Journal of Rural Studies*, 18: 213–217.

Kelly, A. B. and Kowalyszyn, M. (2002) 'The association of alcohol and family problems in a remote indigenous Australian community', *Addictive Behaviour*, 28: 761–767.

Kneale, J. (2001) 'The place of drink: temperance and the public, 1956–1914', *Social and Cultural Geography*, 2(1): 43–59.

Komro, K. A., Maldonado-Molina, M. M., Tobler, A. L., Bonds, J. R. and Muller, K. E. (2007) 'Effects of home access and availability of alcohol on young adolescent's alcohol use', *Addiction*, 102: 1597–1608.

Kraack, A. and Kenway, J. (2002) 'Place, time and stigmatised youth identities: bad boys in paradise', *Journal of Rural Affairs*, 18: 145–155.

Kraftl, P. (2013) 'Beyond "voice", beyond "agency", beyond "politics"? Hybrid childhoods and some critical reflections on children's emotional geographies', *Emotion, Space and Society*, 9: 13–23.

Kurtz, Z. (1999) 'The health care system and children', in Tunstill, J. (ed.) *Children and the State: Whose Problems?* London: Cassell. pp. 92–117.

Kypri, K., Dean, J. and Stojanovski, E. (2007) 'Parent attitudes on the supply of alcohol to minors', *Drug and Alcohol Review*, 26: 41–47.

Langer, B. (2004) 'The business of branded enchantment: ambivalence and disjuncture in the global children's culture industry', *Journal of Consumer Culture*, 4: 251–277.

Leyshon, M. (2008a) 'The betweeness of being a rural youth: inclusive and exclusive lifestyles', *Social and Cultural Geography*, 9(1): 1–26.

Leyshon, M. (2008b) '"We're stuck in the corner": young women, embodiment and drinking in the countryside', *Drugs: Education, Prevention and Policy*, 15(3): 267–289.

Leyshon, M. (2005) 'No place for a girl: rural youth pubs and the performance of masculinity', in Little, J. and Morris, C. (eds) *Critical Studies in Rural Gender Issues*, Aldershot: Ashgate. pp. 104–122.

Lieb, R., Merikangas, K. R., Hofler, M., Pfister, H., Isensee, B. and Wittchen, H. U. (2002) 'Parental alcohol use disorders and alcohol use and disorders in offspring: a community study', *Psychological Medicine*, 32: 63–78.

Longhurst, R. (2001) *Bodies: Exploring Fluid Boundaries*, London: Routledge.

Lowe, G., Foxcroft, D. R. and Sibley, D. (1993) *Adolescent Drinking and Family Life*, Reading: Harwood.

MacKintosh, A. M., Hastings, G., Hughes, K., Wheeler, C., Watson, J. and Inglis, J. (1997) 'Adolescent drinking – the role of designer drinks', *Health Education*, 97(6): 213–224.

Marquis, G. (2004) 'Alcohol and the family in Canada', *Journal of Family History*, 29(3): 308–327.

Martens, L., Southerton, D. and Scott, S. (2004) 'Bringing children (and parents) into the sociology of consumption: towards a theoretical and empirical agenda', *Journal of Consumer Culture*, 4: 115–182.

McIntosh, J., MacDonald, F. and McKeganey, N. (2008) 'Pre-teenage children's experiences with alcohol', *Children and Society*, 22: 3–15.

McKeganey, N., Bernard, M. and McIntosh, J. (2002) 'Paying the price for their parent's addiction: meeting the needs of children of drug-using parents', *Drugs: Education, Prevention and Policy*, 9(3): 233–246.

Measham, F. and Brain, K. (2005) '"Binge" drinking, British alcohol policy and the new culture of intoxication', *Crime, Media and Culture*, 3: 262–283.

Measham, F. (2006) 'The new policy mix: alcohol harm minimisation and determined drunkenness in contemporary society', *International Journal of Drug Policy*, 17: 258–68.

Moisio, R., Arnould, E. J. and Price, L. L. (2004) 'Between mothers and markets: constructing family identity through homemade food', *Journal of Consumer Culture*, 4: 361–384.

Morgan, D. H. J. (1996) *Family Connections: An Introduction to Family Studies*, Cambridge: Polity Press.

Nairn, K., Higgens, J., Thompson, B. Anderson, M. and Fu, N. (2006) '"It's just like the teenage stereotype, you go out and drink and stuff": hearing from young people who don't drink', *Journal of Youth Studies*, 9: 287–304.

Newburn, T. and Shiner, M. (2001) *Teenage Kicks? Young People and Alcohol: A Review of the Literature*, York: Joseph Rowntree Foundation.

Noll, R., Zucker, R. and Greenberg, G. (1990) 'Identification of alcohol by smell amongst pre-schoolers', *Child Development*, 61: 1520–1527.

Ogilvie, D., Gruer, L. and Haw, S. (2005) 'Young people's access to tobacco, alcohol and other drugs', *British Medical Journal*, 331: 393–396.

ONS (Office for National Statistics) (2004) *Census 2001: National report for England and Wales Part 2* (Laid before Parliament pursuant to Section 4[1] Census Act 1920). London: The Stationery Office.

Plant, M. and Miller, M. (2007) 'Being taught to drink: UK teenagers' experience', in Järvinen, M. and Room, M. (eds) *Youth Drinking Cultures: European Experiences*, Aldershot: Ashgate. pp. 20–31.

Plant, M. (1997) *Women and Alcohol: Contemporary and Historical Perspectives*, London: Free Association Books.

Plummer, K. (2003) *Intimate Citizenship: Private Decisions and Public Dialogues*, Seattle: University of Washington Press.

Popham, P. (2005) 'When in Rome, do as young Romans do: binge like a Brit', *The Independent*, 11ᵗʰ August. p. 7.

Postman, N. (1982) *The Disappearance of Childhood*, New York: Delacourt Press.

Radley, A. (1995) 'The elusory body and social construtionist theory', *Body and Society*, 1: 3–23.

Raskin-White, H. (1991) 'Learning to drink: familial, peer and media influences', in Pittman, D. and Raskin-White, H. (eds) *Society, Culture and Drinking Patterns Re-Examined*, New Brunswick, NJ: Rutgers Centre for Alcohol Studies.

Ritson, B. (1975) 'Review: children and alcohol', *Child: Care, Health and Development*, 1: 263–269.

Rolamdo, S., Beccaria, F., Tigerstedt, C. and Torronen, J. (2012) 'First drink: what does it mean? The alcohol socialisation process in different drinking cultures', *Drugs: Education, Prevention and Policy*, 19(3): 201–212.

Rose, G. (1997) 'Situating knowledges: positionality, reflexivities and other tactics', *Progress in Human Geography*, 21: 305–20.

Ruckenstein, M. (2010) 'Time scales of consumption: children, money and transactional orders', *Journal of Consumer Culture*, 10: 383–404.

Shucksmith, J., Glendinning, A. and Hendry, L. (1997) 'Adolescent drinking behaviour and the role family life: a Scottish Perspective', *Journal of Adolescence*, 20, 85–101.

Smith, L. and Foxcroft, D. (2009) *Drinking in the UK: An Exploration of Trends*, York: Joseph Rowntree Foundation.

Templeton, L., Novack, C. and Wall, S. (2011) 'Young people's views on services to help them deal with parental substance misuse', *Drugs: Education, Prevention and Policy*, 18(3): 179–186.

Thrift, N. (2004) 'Intensities of feeling towards a spatial politics of affect', *Geografiska Annaler Series B Human Geography*, 86 B(1): 57–78.

Tyler, M. (2009) 'Growing customers: sales-service work in the children's culture industries', *Journal of Consumer Culture*, 9: 55–76.

Valentine, G., Jayne, M. and Gould, M. I. (2013) 'The proximity effect: the role of affective space of family life in shaping children's knowledge about alcohol and its social and health implications', *Childhood*, 21(1): 103–118.

Valentine, G., Jayne, M. and Gould, M. (2012) "Do as I say, not as I do': family life and the transmission of drinking cultures', *Environment and Planning A*, 44: 776–792.

Valentine, G., Jayne, M., Gould, M. I. and Keenan, J. (2010a) *Family Life and Alcohol Consumption: A Study of the Transmission of Drinking Practices*, York: Joseph Rowntree Foundation.

Valentine, G., Holloway, S. L. and Jayne, M. (2010b) 'Generational patterns of alcohol consumption: continuity and change', *Health and Place*, 16: 916–925.

Valentine, G., Holloway, S. L., Jayne, M. and Knell, C. (2007a) *Drinking Places: Where People Drink and Why*, York: Joseph Rowntree Foundation.

Valentine, G., Holloway, S. L., Knell, C. and Jayne, M. (2007b) 'Drinking places: young people and cultures of alcohol consumption in rural environments', *Journal of Rural Studies*, 24: 28–40.

Valentine, G. (2004) *Public Space and the Culture of Childhood*, Aldershot: Ashgate

Valentine, G. (2002) 'People like us: negotiating sameness and difference in the research process', in Moss, P. (ed.) *Feminist Geography in Practice: Research and Methods*, Oxford: Blackwell. pp. 116–126.

van der Vorst, H., Engels, R., Meeus, W., Deković, M. and van Leeuwe, J. (2005) 'The role of aclohol-specific socialisation in adolescents drinking behaviour', *Addiction*, 100: 1464–1467.

van Zundert, R. M. P., Van der Vorst, H. L., Vermulst, A. and Engels, R. (2006) 'Pathways to alcohol use among Dutch students in regular education and education for adolescents with behavioral problems: the role of parental alcohol use, general parenting practices and alcohol-specific parenting practices', *Journal of Family Psychology*, 20: 456–457.

Velleman, R. (2009) *How Do Children and Young People Learn About Alcohol: A Major Review of Literature*, York: Joseph Rowntree Foundation.

Venn, C. (2010) 'Individuation, relationality, affect: rethinking the human in relation to the living', *Body and Society*, 16(1): 129–161.

Vygotsky, L. S. (1978) *Mind and Society: The Development of Higher Psychological Processes*, Cambridge, MA: Harvard University Press.

Waitt, G., Jessop, L. and Gormon-Murray, A. (2011) 'The guys in there just expect to be laid': embodied and gendered socio-spatial practices of a 'night out' in Wollongong, Australia', *Gender, Place and Culture*, 18(2): 255–275.

Walters, G. D. (2001) 'Behavior genetic research on gambling and problem gambling: a preliminary meta-analysis of available data', *Journal of Gambling Studies*, 17: 255–71.

Ward, B., Snow, P. and Aroni, R. (2010) 'Children's alcohol initiation: an analytical overview', *Drugs: Education, Prevention and Policy*, 17(3): 270–277.

Ward, B. and Snow, G. (2010) 'Supporting parents to reduce the misuse of alcohol by young people', *Drugs: Education, Prevention and Policy*, 17(6): 718–731.

Webb, J. A., Baer, P., Getz, J. G. and McKelvey, R. S. (1996) 'Do fifth graders' attitudes and intentions towards alcohol use predict seventh-grade use?' *Journal of the American Academy of Child and Adolescent Psychiatry*, 35(12): 1611–1617.

Wells, K. (2002) 'Reconfiguring the radical other: urban children's consumption practices and the nature culture/divide', *Journal of Consumer Culture*, 2: 291–315.

Wickama, K., Conger, R., Wallace, L. and Elder, G. (1999) 'The intergenerational transmission of health-risk behaviours: adolescent lifestyles and gender moderating effects', *Journal of Health and Behaviour*, 40: 258–272.

Yu, J. (2003) 'The association between parental alcohol-related behaviours and children's drinking', *Drug and Alcohol Dependence*, 69: 253–262.

Index

For Product Safety Concerns and Information please contact our EU
representative GPSR@taylorandfrancis.com
Taylor & Francis Verlag GmbH, Kaufingerstraße 24, 80331 München, Germany

9 780367 219055